Georg F. L. Stromeyer, Friedrich von Esmarch, Sherrard Freeman Statham

# On the fractures of bones occurring in gun-shot injuries

Georg F. L. Stromeyer, Friedrich von Esmarch, Sherrard Freeman Statham

**On the fractures of bones occurring in gun-shot injuries**

ISBN/EAN: 9783337159795

Hergestellt in Europa, USA, Kanada, Australien, Japan

Cover: Foto ©berggeist007 / pixelio.de

Weitere Bücher finden Sie auf **www.hansebooks.com**

# GUN-SHOT FRACTURES,

BY DR. STROMEYER,

AND

# RESECTION IN GUN-SHOT INJURIES

BY DR. ESMARCH,

*(Slightly Abridged.)*

TRANSLATED BY S. F. STATHAM,

WITH SOME

## RESECTIONS ON TONIC TREATMENT,

BY THE TRANSLATOR.

*ILLUSTRATIONS OF INSTRUMENTS USED BY THE SAME.*

ON THE

# FRACTURES OF BONES

OCCURRING IN

# GUN-SHOT INJURIES.

BY

DR. LOUIS STROMEYER.

---

ON

# RESECTION IN GUN-SHOT INJURIES.

OBSERVATIONS AND EXPERIENCE

IN THE

SCHLESWIG-HOLSTEIN CAMPAIGNS OF 1848 TO 1851.

BY

DR. FRIEDRICH ESMARCH.

(SLIGHTLY ABRIDGED).

---

# CASES OF RESECTION IN CIVIL PRACTICE.

ON

TONIC TREATMENT THROUGHOUT.

BY

S. F. STATHAM.

LONDON:

TRÜBNER & CO., 60, PATERNOSTER ROW.

1860.

# PREFACE

OF THE

## TRANSLATOR OF THE FIRST TWO PARTS OF THIS WORK.

---

THE results of the experience of Dr. Stromeyer in Schleswig-Holstein are the reasons for translating this portion of his "Handbook of Surgery."

These results especially refer to the resection of the elbow, and, again, of the shoulder; the first practically introduced by Dr. G. B. Langenbeck, and both of them proved by the same surgeon and by Dr. Stromeyer, to be far more universally applicable in military surgery, than had been formely supposed.

Dr. Stromeyer's success in the cure of comminuted gun-shot fractures of the shafts of all the long bones is especially worthy of remark.

Dr. Esmarch's Treatise merits particular notice, on account of the results given in figures, both in the text and at the end, in a tabular form.

I cannot at all agree with the antiphlogistic treatment so strongly recommended, indeed, while in Schleswig-Holstein, I began to treat the patients on the opposite plan—and successfully so.

To any one interested on this point, and every surgeon must be so, I believe my own cases of resection, at the end of the volume will be of great interest.

S. F. STATHAM.

*April*, 1856,

43, *Mortimer Street, Cavendish Square.*

# CONTENTS OF FIRST TRANSLATION,

## *(Dr. Stromeyer.)*



# CONTENTS OF SECOND TRANSLATION,

## *(Dr. Esmarch.)*

## CASES IN CIVIL PRACTICE,

*(S. F Statham)*

ON THE

# FRACTURES OF BONES

OCCURRING IN

# GUN-SHOT INJURIES.

BY

DR. LOUIS STROMEYER.

[BEING THE FIFTH PART OF THE HANDBOOK OF SURGERY OF THIS AUTHOR.]

FREIBURG IN BREIGAU.

HERDER.

1850

# ON THE FRACTURES OF BONES

OCCURRING IN

# GUN-SHOT INJURIES.

---

The fractures occurring in gun-shot wounds offer many peculiarities in their course by which they are distinguished from the ordinary complicated fractures as recognised by the practical surgeon. A special consideration of these fractures may be so much the more suitable here, as I have had occasion to observe more than two thousand gun-shot wounds, as Surgeon-in-Chief of the Schleswig-Holstein army, in the campaign of 1849 against the Danes.

The most fearful injuries of this kind are those caused by the heavier projectiles,—cannon-balls, pieces of bombs, fragments of wood or of other material which, torn loose and scattered around, cause comminuted fractures of the bones, and in the soft parts a concussion which diminishes their vitality, so that the blood stagnates in the crushed vessels, and the innervation being in a decreased extent, gangrene commences. Very frequently these injuries are fatal from the shock experienced by the nervous system, showing itself by an extreme degree of weakness, coldness and pallor, and by a rapid, small pulse; often death occurs in the first twenty-four hours, should the patient live longer gangrene sets in, which generally attacks all parts below the affected spot, and now the patient rarely survives the fourth day, remaining in a depressed condition with a more rapid, but by no means strongly developed pulse. This mode of death occurs most frequently in those persons where a heavy projectile has torn away a whole limb, often they die soon after the reception of the injury, many indeed on the field of battle; the least dangerous is the loss of an arm, and again in such cases the danger is increased the nearer the injury is to the trunk. Of four cases of this kind which I have seen, only one recovered where the arm was torn away in the elbow-joint, the three others—where the injury took place above the insertion of the deltoid—were fatal; in one of them exarticulation was performed at the shoulder and proved fatal within the first three days. A fifth case, in which the arm was torn away above the elbow by a cannon ball, and in which amputation was soon after performed, proved fatal on account of rupture of the spleen, with fracture of the ribs and injury of the lung, caused by the arm being violently struck against the side; the patient lived to the eighth day. A similar case ended fatally after complete healing of the stump, with symptoms of suppuration of the kidneys, their rupture having been indicated in the first instance by obstinate vomiting, ice-coldness of the extremities and hæmaturia. In these fatal cases after the arm has been carried away, the simultaneous injury to the trunk must be always taken into consideration, the risk of this is less in similar accidents in the lower extremities; while there, as well known, the danger of amputation is greater.

Not less dangerous than the cases last mentioned are those where the bone is crushed by a heavy projectile yet the soft parts are preserved, these are very apt to result in mortification. By no means seldom, the appearance of the limb deceives one as to the extent of injury, so that it is supposed a simple fracture is present, yet upon closer examination extensive crushing is discovered. If such wounded persons must be far removed, they are lost almost without exception and are not to be saved by amputation: is no removal necessary, the result of amputation is more favourable. The extent of the injury is recognised by the peculiar feeling of many fragments of bone rubbing upon one another, by the increased mobility, and by introduction of the finger, if a wound is present. But one example is come before me where simple fracture of the leg was caused by a cannon-ball, and this was by means of the metallic scabbard being struck and driven against the leg. In a similar case the upper fragment of the tibia had pierced the soft parts.

Wounds by case-shot hardly occurred in the last campaign, an example which I observed in Freiburg was undistinguishable from a musket-wound. By far the great majority of gun-shot wounds are those caused by musket- or rifle-bullets, with which must be reckoned the 'Espignole,' a species of fire-arm employed by the Danes, consisting of a long barrel from which a number of bullets are discharged one after another and which as the jet of a fire-engine may be directed to various points. The bullets are conical and perforated in their long axis, this perforation conducts the ignition from before to the powder laying behind. Of these bullets there are two varieties the smaller about the weight of a rifle-bullet, the larger double the size, their action corresponds with that of musket-bullets. The musket- and rifle-bullets are divided into globular and conical bullets, the last may be employed with the ordinary or with rifled barrels. It has been supposed that the conical bullets cause greater destruction of the bones than the ordinary globular ones, however I have not been able to assure myself of this, their action appears to be entirely equivalent; again it is by no means the fact, that the kind of bullet can be recognised by the perforation made by it in the clothing, it was thought that the conical bullet made a T-shaped opening, and that the ordinary bullet carried away a round piece, not seldom however is this action reversed. After the battle of Fridericia I removed by incision a conical bullet from the fossa infraspinata of an Officer, where the bullet had pierced the pectoralis major and perforated the scapula from within outwards, it was still wrapped up in cloth and padding which it had torn away. A much lamented Officer was shot in the head, I found in the leather shade of his cap a T-shaped opening, his friends supposed from this circumstance that he had been struck by a conical bullet, however in the autopsy a globular one was found.

The action of Bullets on the bones are the following;

1. They strike the bone without breaking it and flatten themselves against its surface; the bone struck becomes necrotic from the destruction of its periosteum. In crowded hospitals such injuries of the larger long bones cause suppuration of the medullary canal, which extending itself, at last by the

passage of pus into the veins, gives a fatal termination. In the autopsy (the bone being sawn in its long axis,) the marrow is found filled with pus from the wounded part upwards and the same morbid product in the neighbouring large veins; as in the femoral vein after contusion of the femur.

The spot struck by the ball is colourless and exsanguine, in its circumference appears the commencement of a line of demarcation. Contusions of this kind which, up to the present time have been little attended to in the long bones. are well known in the bones of the skull, where caused by a blow or fall they have the same dangerous consequences if not properly treated, as suppuration occurs in the diploe and purulent inflammation in the sinus' with its usual results. Such contusions also occur in the cranial bones, if a bullet strikes at a right angle, of which I have seen many examples, where it could be determined from the character of the wound in the soft parts, that the same had been so struck, without causing fracture or depression of the bone. Usually however such contusions of the cranial bones arise from grazing shots, in which the bullet strikes in a more or less obtuse angle, pursues its course for a certain distance on the skull, partly tears off, partly bruises the pericranium and leaves behind one or more slight depressions of the outer table. These depressions often lie at some little distance from one-another, so that it seems as if the bullet had sprung over a part of the skull: this it might easily do if striking at an obtuse angle it should cause a slight turning of the head, rebound, and immediately after strike the skull at some other point. I have a preparation which shows this action of the bullet very plainly. In these cases the tabula vitrea is generally splintered; upon this, see Injuries of the Head.

2. The bullet breaks the bone, striking it at a right or obtuse angle, and to the side of its shaft, it is now turned out of its straight direction by the bone so that the apertures of entrance and exit do not seem to correspond, yet certain cases occur where the apertures are so situated, that it is evident the bullet must have passed through the space which the bone occupies, and nevertheless the fracture is only simple, without comminution; hence it must be considered that the breaking bone had given way to the bullet. Such cases under otherwise favorable circumstances end often very successfully, even in such largo bones as the femur. This kind of injury is recognised by the more lateral direction of the course of the bullet as regards the bone and by the crepitation as in ordinary fractures.

3. The bullet impinges upon a long bone laterally and crushes a part of its cylinder, without entirely dividing its continuity; I possess a preparation from a femur in which close above the condyles in a length of two inches, the half of its cylinder was struck out by a bullet, which I removed the day of the battle of Fridericia together with a portion of the splinters of bone,—the bullet had entered anteriorly and had lodged behind the fragments. Those cases are similar where a grazing bullet has torn up a groove in the bone, as I have observed it on the posterior surface of the condyles of the femur, without important injury to neighbouring parts and with successful issue.

4. The bullet is driven into the bone and remains fixed, without splintering it further or causing fissures. This takes place most readily in the spongy bones of the foot, in the pelvic bones and upper third of the tibia.

5. The bullet pierces the bone and forms a canal without causing further splintering. This occurs in the cranial and pelvic bones, in those of the hand and

in the upper third of the tibia. As the soft parts in this species of cases are at the same time much injured, severe inflammatory symptoms follow, these cases are often however favorable, as I have twice observed in the tibia, which had been pierced from before backwards by a bullet.

6. The bullet bruises the bone with or without sensible depression, or becomes seated in or traverses it, at the same time however gives rise to fissures into the the next articulation. I have observed this case most frequently in the elbow- and knee-joint, in the last named both in the femur and tibia. The diagnosis here is extremely difficult, indeed in many cases it cannot be formed with certainty, as the fissures in question are so fine, that they may neither be felt by the finger nor by a probe. In the first instance the motion of the joint is perfect, it is when the suppuration from the wound becomes considerable that the joint inflames and likewise proceeds to suppuration in spite of all antiphlogistic means. After the battle of Colding we had in Christiansfelde four such implications of the knee-joint, in which it was doubtful for many weeks whether the limb could be saved, or not. In three of these cases amputation became necessary, and in each fissures into the joint were discovered in the removed limb. In the fourth the limb was preserved, and here doubtless the fissuring had not taken place. But one case is come before me, where the tibia being struck on its posterior surface close above the ancle-joint and the fibula being grazed, fissures extending to the joint were indicated, as it became much swollen and was the seat of very severe pain. A cure nevertheless ensued with anchylosis, and to this certainly the use of ice, and internally of large doses of Opium much contributed.

7. The bullet strikes a long bone at a right angle in its shaft and forces its way through it, breaking the part into many pieces, the fissures passing more or less upwards and downwards according to the long axis, for a considerable distance from the affected spot. This happened most frequently in the tibia, but as the fibula often remains uninjured, the dangerous condition is not readily known, here also the fissures not being recognisable. The loose splinters are removed and it is hoped that under suitable treatment the limb may be preserved; in spite of this, however, a widely diffused inflammation sets in, proceeds to suppuration and continually extends itself unchecked by incisions. In certain cases a cure follows after exfoliation of extensive secondary and tertiary sequestra,—in others amputation becomes necessary from the extensive suppuration, or from parenchymatous bleeding, still under such circumstances, the life is rarely preserved.

The parenchymatous hæmorrhages are, as I have found, and will elsewhere prove by facts, at once a symptom of the entrance of pus into the veins, and of stoppage of the larger veins by coagulation. The stagnation of blood thereby ensuing, gives rise to hæmorrhage from the capillary vessels lying free in the wound and on this account, the blood so lost has neither a decidedly venous nor arterial character. Nevertheless I will not assert that parenchymatous hæmorrhages—independent of stagnation in the veins and similar to scorbutic bleedings—are not met with from the surface of wounds.

### On the Examination of Gun-shot Wounds, the Bone being injured.

It may be believed that nothing is more easy to diagnose than an injury of bone, with a wound extending to it which offers no hindrance to the introduction of the finger. Indeed one might think that these fractures must be much easier to recognise than simple fractures. However experience has taught me that in this respect more errors than right diagnoses are made. The cause of this is, that the majority of surgeons employ for the examination of gun-shot wounds— not their fingers— but the probe, with which in fact nothing can be properly felt. It would be by no means improper, to deprive military surgeons of their probes in the commencement of the campaign, in order to lead them to use their fingers for probing wounds. I do not remember having used any other probe than the director [ German, grooved probe ] during the whole campaign. Another difficulty exists in the character itself of these fractures,—for while in ordinary fractures a complete division of the periosteum takes place, this is frequently preserved in its greater extent in gun-shot wounds, and often binds together the fragments so thoroughly, that no displacement of them follows and the form of the bone is retained. Thus I have seen complete crushing of the upper end of the ulna into the joint, in which no deformity occurred and the patient could perform all the usual movements. Even in examination by the finger it is easy to be deceived, if the examination is not conducted in the very spot where the shot had entered, and from the bullet changing its course after striking the bone, the finger may follow the track formed in the soft parts without attending to the injured bone. The longer the track is, the more difficult are these examinations, hence many injuries of the bones are first known by their consequences.

Prognosis in Gun-shot Injuries of Bones.

It cannot be forgotten that gun-shot wounds accompanied with injury to the bone, are much more dangerous than those not complicated in this manner. Indeed the first effect of the wound is much more severe if the bone has been struck,— for when a bullet merely pierces the soft parts, very often the flowing blood first shows the patient that he is wounded,— should however a bone be struck, the patient is thrown to the ground, as it were torn away from the place or carried round in a circle; while the usual phœnomena appear of a shock to the whole nervous system viz; syncope, coldness, paller and a small, quick pulse persisting for a longer or shorter time after the reception of the injury. This shock is probably not without influence on the further course of the case.

By a greater size of the injured bone, the more extensive is its injury and the nearer this is to the trunk, so much the more guarded must be the prognosis. The irritation of the soft parts by the broken ends of the bone is of the greatest influence on the subsequent inflammation, so that there is an important difference in the course of the case should one bone be broken or both of them, — while in the latter the most serious inflammatory symptoms occur— the former runs its course with merely local symptoms, except in such cases where the splintering extends very far. These observations are of the utmost importance for the treatment of such cases, as we see that the injury to the bone is not the cause of the violent inflammation, but that this arises from the irritation of the soft parts by the sharp fragments of bone thrust out of their position either by external violence or by the muscles of the limb.

For treatment of individual cases of this kind, these remarks remain in their full force, but in hospitals where the pyæmic contagion developes itself, the injured bones experience a far greater danger— one entirely independent of the irritation caused by the fragments. Experience has taught me, that it is especially those gun-shot wounds with injury to the bone which end fatally by pyæmia. The nearer to the trunk the bones of the extremities are injured so much greater is the danger,—in the trunk itself the injuries of the pelvic bones are especially combined with risk of the occurrence of this complication.

The cause of this great danger in gun-shot wounds of the bones is evidently due to the fact that such wounds become cleansed much slower than wounds merely of the soft parts, which suppurate kindly and contract themselves after the removal of the foreign body,—while in injury to the bones the throwing off of the sequestra may endure for months. Besides the structure of the bone itself is a cause of the easy introduction of acrid matter into the veins, as these do not in bone as in soft parts collapse after injury, but their cut extremities remaining open are exposed to the entrance of pus. The suction of the venous blood by means of the respiration exerts its action also upon the fluid bathing the wounded bone. Thus the pus or rather the serum of the pus passes deeper and deeper into the spongy substance or into the medullary canal, setting up here fresh suppuration and not unfrequently reaching the extremity of the bone. I have many times convinced myself, that the pus first passes into the larger veins out of the osseous substance and thence reaches the general circulation. In those cases of amputations of the thigh ending fatally through pyæmia—mostly after the lapse of many weeks,— the crural vein at its divided extremity was thoroughly closed and for a considerable space from thence filled with a firm clot, and it was only higher above where the veins returning from the bones enter, that we found the femoral vein containing purulent fluid. The purulent phlebitis therefore had not begun at the wound or the vein would not have become closed, and where so large a vein had fairly healed it is not to be expected that smaller veins remaining open could have conducted the sanious fluid or a suppurative inflammation from the wound. It must also be remembered that here on the coast of the Baltic, the spontaneous, rheumatic, suppurating and pernicious inflammations of bone are very common. This circumstance must have an influence in the class of injuries we are considering.

Should these principles be correct, it would be of the greatest importance to consider means by which the medullary canal could be withdrawn from the action of the purulent fluid, whether by cauterisation or by mechanical measures. The discovery of the proper method must be left to the future. The best of course would be, to prevent in every possible manner the development of a contagious pyæmia; this is so much the more necessary, as frequently the wounded bones are withdrawn from all direct interference without the danger as regards pyæmia being in any degree lessened. This occurs for instance in numerous injuries of bones which cannot be recognised during life, as in many injuries of the pelvis, in cases of gun-shot wounds of the upper part of the thigh. If in such cases a rapid pyæmia set in, I was not mistaken in diagnosing an injury of bone, and the autopsy confirmed this by the presence of a contusion, fissure or fracture. After the battle of Fridericia I extracted a bullet from the middle of the deltoid of the left arm, which had entered close above its insertion. On the eighth day pleurisy of the left side occurred, and the patient sunk rapidly with the usual symptoms of pyæmia; on the examination a depression was found close beneath the tuberculum majus of the humerus, whence some fissures proceeded upwards and downwards. During life there were no signs of inflammation of the shoulder-joint, nor of excessive inflammation of the wound. It is not to be wondered at that in the trunk or its neighbourhood, the injuries of bone lead more readily to poisoning of the blood as here the influence of inspiration in drawing the pus into the veins is stronger than in parts further removed from the trunk. For the rest, there are constitutions which have no susceptibility of the occurrence of pyæmia, and in whom the excessive suppuration alone at last causes death. In this way I saw two cases of crushing of the neck of the femur slowly terminate, where the quicker death by pyæmia would have been a blessing,—on the autopsies all the internal organs were sound.

## Course and Treatment.

The military surgeon, Dr. N. N. said once in relation to the result of gun-shot wounds: "those cases proceed the best, which are not meddled with." It would therefore appear that the haste of the surgeon to do something, is not always of service to his patient.

The state of a fresh wound is by no means rendered worse by the introduction of the finger:—to assure oneself of the extent of injury, to extract the foreign body, entirely separated pieces of bone or portions of clothes forced in with the bullet. A dilatation of the wound for this object must be employed when necessary and generally in the direction of the long axis and may be useful to withdraw the bullet, if its removal is possible, its position being known and the attempts not becoming tedious by their frequent renewal. It always is the best practice, when the wound has been so far dilated, that the bullet can be reached by the left index finger, upon this to introduce the ordinary dressing forceps; as soon as the forceps touch the foreign body the finger is withdrawn somewhat to allow the forceps to be prudently opened, and in this state they are passed over the bullet, so that thus it is readily and safely removed. The objections to the use of probes have been already mentioned.

Of all the foreign bodies that are found in a gun-shot wound, the bullet is clearly the least hurtful, as shown by the frequent cures with enclosure of the bullet, even in such cases where injury to the bone was also present and suppuration had on that account endured for a very long time. Hence the continued search for a bullet is certainly not to be justified;—can it not be felt from the wound, or discovered and removed by incision on the opposite side of the limb, it should for the time be left uninterfered with.

This remark, which is true as regards the bullet, is also applicable to sequestra; those only should be removed which are fully separated and easy to reach. It is wonderful with what trifling exfoliation fractures are frequently healed, in which no attempt has been made for the extraction of splinters; such cases are generally those where the bullet has traversed the limb, hence giving no occasion to search for it. Above all, the removal of splinters still hanging to the periosteum is to be guarded against,—should these be separated by the subsequent inflammation, the periosteum still remains behind and forms new bone, while at the first the splinters could only be removed WITH the periosteum.

Acting upon this view, I considered extensive resections in the continuity of bones as unsuitable, as in recent cases this can only be performed with a corresponding loss of periosteum, and therefore leads to the danger of non-union of the resected extremities. This operation is otherwise dangerous from its effect on the whole system, and so much the more so, according as the wound is less recent, and again, as inflammation or even suppuration has further developed itself. I do not reckon with these resections, the removal by a small saw or by bone-forceps of sharp points of bone in recent wounds, thisisa practice I have frequently adopted, here there is no great exposure by dilatation, nor is the contour of the injury of the bone altered essentially.

I have had no cause to regret this abstinence from resection in the continuity, on the contrary I have won the assurance, that it is for the most part an unnecessary, and otherwise from its danger, a prejudicial operation. Those resections that I have seen practised in the proper continuity, have ended fatally; in a resection of the radius not proving fatal, where two inches were removed by the saw, no union occurred, the hand was drawn entirely to the radial side and continued limited in its functions, while in all similar cases without resection union was perfect. At the same time in cases where no resection has been practised and where the splinters have been loosened by suppuration, shortening occasionally results from the periosteum having been destroyed by the bullet or by the suppuration itself. In a Saxon officer, I removed on the ninth day, the fragments loosened by suppuration of a fractured radius, during the progress of cure the hand was drawn towards the radial side, but the bone became firmly united and in spite of a slight obliquity of the hand its use was not otherwise interfered with. Again though no resection has been employed false anchylosis may result, should the loss of substance have been great,—this I have observed in the lower maxillary bone and in the metacarpus.

What I have here said on the cautious extraction of foreign bodies from fresh wounds applies in double force to wounds already inflamed where the entrance of the track is so swollen as scarcely to allow of the introduction of the index-finger and where the patient on account of the commencement of the inflammation is become very much opposed to any examination. In such a state an examination should only be made when the question is, whether an important operation should be performed, or not. Is the limb to be saved, the wound as a rule should neither be disturbed by the finger nor by probes. In fresh wounds the extraction of foreign bodies and the dilatation of the wound possibly requisite for that purpose, may serve to remove a part of the blood effused in the neighbourhood of the fracture, but after the first twenty-four hours this is no longer the case. An operation would now admit air more freely to the effused blood, which decomposing becomes incapable of absorption. Thus it excites suppuration, essentially contributes to enhance the inflammatory symptoms and thereby to induce necrosis of the fractured ends;— while in a less degree of inflammation the partially loosened fragments alone become separated and the ends of the broken bone exfoliate imperceptibly, just as we sometimes observe in ordinary complicated fractures.

Dupuytren distinguishes very suitably in the gun-shot injuries of bones the Primary, Secondary and Tertiary Splinters. The Primary are those found fully loose and which are drawn out on the first examination. Secondary,— those still hanging to the soft parts and becoming free by suppuration, how soon these can be withdrawn depends upon their position, if superficial their discharge may take place without any violence from the fifth day, if deep or very large it may be many weeks before this is possible. The degree of reaction has also great influence on this point: should the inflammation be kept under by rest, abstraction of blood and the application of cold—the loosening of the fragments is much hindered; on the other hand, it proceeds rapidly if poultices have been employed without any antiphlogistic means.

Tertiary splinters, are those which arise through inflammatory necrosis of the fractured ends. The larger these sequestra are, so much longer time is requisite for their separation; it is these especially which delay the recovery. They lie in the wound as foreign bodies, keeping up suppuration and yet cannot be extracted. The more profuse this suppuration is, so much less hope is there of reproduction of bone, and often the fragments only unite if by overlapping they come in contact with the sides—instead of with the necrosed ends—of the bones. Thus union is accompanied by considerable shortening and great deformity. In cases where the necrosis proceeds very deeply into the fractured ends, so that for a certain length the whole diameter of the bone has died, the cure may sometimes fail from the difficulty of removing the sequestra, which as in other cases of necroses, are sometimes enclosed in new formations of bone. This necrosis however does not always proceed deeply, but may only affect a very thin layer, so that the sequestrum is discharged in small particles and the limb preserved, while necroses of larger extent may require amputation at any future period.

On account of the great influence which the necrosis of the fractured ends exerts in interrupting the union, it must be our most earnest effort to put a stop to this process. This is not always in the power of the surgeon; the shattering of the bone and the simultaneous injury to the soft parts—especially should vessels and nerves be crushed,—leads without fail to violent inflammation. A chip of a bullet, a piece of the clothes may be between the fragments and cause obstinate suppuration and necrosis.

Treatment.

The most powerful means for obviating inflammatory necrosis are, Rest, Antiphlogistic Treatment and Care for the free escape of matter.

The required Rest is to be obtained by suitable position and retention as in ordinary fractures, so that only the attendance on the wound demands special care. Where practicable the part should be placed in such a manner that no change of position is requisite in dressing the wound; nothing is more prejudicial during the first few days, than the frequent raising of the injured limb. A young officer in whom the humerus, two fingers breadth below its head, was shattered into several pieces and in whose case it had been wished at the first to perform exarticulation at the shoulder-joint, complained bitterly to me of one of our surgeons, because the first dressings had been left so long that maggots had bred in them—I told him however that he must thank this surgeon for the preservation of his arm. The importance of as complete rest as is possible of the wounded limb is unfortunately too little acknowledged; I have often with great discontent been forced to see such limbs withdrawn from their quiet position, raised and held unsteadily, while being washed and dried. Some trouble and attention would render it by no means difficult to avoid this meddling in the majority of cases, for they are very painful to the patient, frequently give rise to hæmorrhages of small amount, and continually kindle the inflammation anew. In what way this is to be carried out must be studied for each limb.

The second capital point is the Antiphlogistic Treatment. A great man, John Hunter, says, in his "Treatise on Gun-shot Wounds," that injuries of the extremities do not bear venesection so well as those of the cavities. The majority of surgeons follow this doctrine; I have already said in reference to ordinary fractures that I consider it erroneous, and that I employ venesection in complicated cases with the best results. From all that I have seen in this campaign, I am well able to hold the same opinion as regards gun-shot fractures. Unfortunately, however, in these days the value of venesection as well as of the most active remedies is sunk through the influence of the present expectative, homœopathic and hydropathic medicine, so that the young German surgeons glory in the fact, that they have never ordered a bloodletting. Because at present many inflammations of the lungs are recovered from without abstraction of blood, it is thought the severest injuries of the head must be similarly handled, or even wounds of the cavities of the head or chest. The small contracted pulse which at first with gun-shot wounds is very frequent, seems besides to forbid opening a vein, yet it very often becomes fuller by such treatment. The proper time for venesection is within the first three days, from the point when he patient completely recovers from the first shock; when once suppuration is present, it is no longer indicated, but now, the local abstraction of blood by leeches and in many cases by incision. As regards the last means, I have found that leeches cause such a diminution of the inflammatory swelling in the neighbourhood of the fracture, that I have always more sparingly employed incisions, and less to induce local abstraction of blood as to favour the escape of matter. For the rest, suppuration is often very tardy in appearing in gun-shot wounds, so that venesection can often be employed after the third day with the greatest advantage. The chief ground why so many surgeons are chary of abstracting blood in cases of complicated fractures is, that they fear in this manner to weaken patients who must later endure extensive suppuration; they do not think that the suppuration will be so much slighter, the more an impression has been made upon the congestive stage of the inflammation. Without reference, however, to the future danger of suppuration, which is opposed by suitable abstraction of blood in the congestive stage, similar help is required by the nearer risks of violent inflammation. I have seen several patients die from injuries of the upper extremities within the first four days, merely from the severity of the inflammatory symptoms together with febrile delirium, before the formation of matter, and with no other cause for the extensive swelling about the wound, and for the fatal termination having been observed, than the entire omission of abstraction of blood. Of course the rarer cases are not here included, where, in the first few days, under general typhoid symptoms, a putrid condition of the wound without preceding inflammation appears, nor those where pyæmia sets in within the first five or six days.

I cannot sufficiently impress upon young military surgeons the importance of not omitting the abstraction of blood; in reference to this, my views agree entirely with those of military surgeons at the time of the French war, particularly of the English surgeons. If, also, the essential nature of diseases are much less inflammatory now than thirty years ago, and this point deserves the closest observation in surgical practice,—yet the experience obtained from internal diseases cannot find unconditional application in the mechanical injuries of healthy and strong young people; a gun-shot wound is entirely different from an inflammation of the lung, on whose course the epidemic and endemic poisons exert great but not proportionate influences.

Cold applications are next in importance to abstraction of blood as an antiphlogistic means. Unfortunately in the campaign of 1849 we had no ice at our disposal; but in Freiburg in 1848 at the time of the revolution I had opportunity to assure myself of the great value of its application. In this form cold can be applied in the gentlest and least annoying manner. In the absence of ice, the application of cold water must sometimes take its place without nearly equalling it in value. Cold is chiefly useful in hindering the progress of the inflammation and the access of suppuration, so that thus it becomes possible for a large quantity of extravasated blood between the muscles at the seat of fracture, to be removed by absorption. Should however inflammation and suppuration set in, the masses of extravasated blood become purulent foci; these, surrounded by inflammatory effusion exert a pressure on the veins and hinder the flow of blood, so that congestion and widely spread serous infiltration ensue. These infiltrations are well known to military surgeons, and are always a subject of fear, without it having been attempted to explain their occurrence. They are very often the forerunners of phlebitis and its fatal termination. I have seen them much more seldom in such hospitals where cold applications were employed in the first instance, than in those where poultices were similarly used.

In warm seasons and in certain patients cold applications are often borne for a long time; they should be left off when they seem less beneficial to the patient, or if they excite pain. Poultices need not always follow;—the cold applications may be allowed to become warm, or lukewarm Goulard's lotion be employed. However, there arise cases where poultices are urgently demanded. especially in wounds of the hand or foot, or of the fore-arm or leg, when full of sharp fragments of bone. Here they promote suppuration of the outer wound and the escape of secretion, and are often extraordinarily beneficial. It is not always necessary to envelope the greater part of the circumference of the limb in a poultice, indeed the necessary lifting up of the limb each time to attain this object is much to be blamed; though many surgeons do it, picturing to themselves that to embed the limb, nothing is equal to soft pap, without thinking that the mere raising of the limb injures it far more than the poultice can benefit it. All that is requisite is, to lay a sufficiently large poultice on the exposed side of the member, this is covered with a piece of oiled cloth, with cotton-wool and flannel, so that it remains warm and does not require frequent renewal. In many cases a dressing of oiled charpie and compresses above properly fastened, may at once follow the use of cold applications, adhesive plasters being avoided when possible.

I cannot allow that the use of poultices in gun-shot wounds of bones is always attended with great danger, and Guthrie's assertion that amputation generally follows the poultice is certainly erroneous. Poultices promote the yielding of the aponeuroses to the inflammatory swelling, the flow of secretions of the wound externally, and by increasing the activity of the skin they favour absorption. Being generally so agreeable to the sick, it is not to be wondered at that many surgeons employ them almost universally. They require great care in their management, this is scarcely possible in the field. Should the patient be once accustomed to them, they are not easily omitted; for dry applications give pain, and hence one is obliged to continue the use of poultices for months. I am therefore of opinion, as already expressed, as regards compound fractures, that poultices should be used in gun-shot wounds with great moderation only, and that they should be omitted as soon as possible, for through their long employment there supervenes a general atony of the injured limb by which it is disposed to œdematous swelling, and the contraction of the cutaneous papillæ becomes impossible, so that the cure does not advance properly. It now becomes necessary to employ flannel-bandages which answer admirably.

The third chief point to be attended to is, for the Free escape of the wound-secretion; not only pus, but also the first serous exudation of inflammation should have free exit. Should these be retained they would later be replaced by suppuration, while none would have occurred could they have escaped freely. This is often very evident in the treatment by incision of inflammations situated beneath fasciæ. Should the incisions be made early, the exudation being yet serous, flows at once from the wound, and afterwards we have merely a trifling suppuration of the incision. If the incision is delayed there results an extensive suppuration often dangerous to life, and the incision then employed is not so useful by far. This experience should also find its application in the treatment of gun-shot wounds. The canal of a shot which has traversed the limb in a straight direction, does not give much exudation at the first, although a bone may be injured. If, however, the bullet glance from the bone and continue its course in another direction, it often tears the muscles apart from each other, forming cavities under the fascia in which the finger loses itself, so that frequently it is impossible to say what course the bullet has actually taken. These cavities fill themselves very quickly with serous exudation, which is so much less able to flow out of the narrow openings, as these soon become smaller from the surrounding swelling.

In all such cases I am of opinion that the entrance opening, and, when necessary, also the issue, should be dilated by dividing the muscles and fascia according to the long axis of the limb with a probe-pointed bistoury upon the introduced finger. The majority of modern surgeons express themselves against the dilatation of fresh gun-shot wounds, unless some positive object requires it, as extraction of foreign bodies or ligature of a wounded vessel. It is with this as with all other active remedies, which first by exaggeration come into discredit, and only tardily again acquire their true position. One soon becomes assured that the enlargement of all gun-shot wounds is an useless cruelty; however, the fault must be avoided of considering them always useless: it will be the duty of the present generation to determine their real value. I have convinced myself during the late campaign, that the omission of dilatation in such cases as have been mentioned has very serious consequences. Extensive suppurative inflammations occur requiring later many incisions and dilatations, which early enlargement of the wound would have obviated. This is true especially of gun-shot wounds of the thigh, where the cavities torn open by the bullet are in the neighbourhood of great vessels. In these cases is also superadded the great danger, that the femoral vein being bathed in pus may inflame, causing death through pyæmia, as I have frequently seen in cases where no injury to the bone was present. In such cases early dilatation of the wound would have prevented the misfortune, by allowing the serous effusion to escape.

Beyond the enlargement of fresh wounds, the other rules for the escape of effusion can be given in a few words. The part is so placed that the matter may as far as possible flow out from the effect of its own gravity;—should the openings present be too small for that purpose they are enlarged;—is matter burrowing—the passages are to be slit up or a counter-opening to be made, and should new formations of pus take place, the abscesses are to be opened as soon as fluctuation is become distinct. Very often it is preferable to precede the formation of pus in such new swellings by the application of leeches or by an incision piercing the fascia. As regards such incisions most modern surgeons are sufficiently enterprising; yet as regards the allowing matter to be discharged, I have found the most prejudicial timidity is the fashion. Instead of favoring the escape of pus by a suitable position of the part, and incisions and bandages to correspond, the wound is shut up immediately with a pledget of smooth charpie and strips of adhesive plaster, above this a compress and bandage, so that frequently nothing can flow away, and on the dressing being removed in twenty-four hours, a full stream of matter rushes out. Now comes the pressing and squeezing around to force out the last drop of matter,—many taking the limb between their hands and squeezing it as they would a lemon. Let this treatment be continued only some days, and the pus, at first mild, scentless and yellow, becomes acrid, greenish and fœtid. As usual, one fault introduces another; a compressing bandage is now made use of; from this I have never seen the least use, but often the greatest disadvantage, as it retains the pus rather than forces it out. When this also proves useless, the next step is to use injections, generally of lunar caustic, sometimes of camomile tea, chlorine water or decoction of cinchona, on purpose to improve the quality of the discharge; this is only become acrid because its escape was not free, and it becomes no better from the presence of such injections, a part of which always remains behind in the wound,—the most harmless indeed are warm water injections, without, however, attaining the desired object. It is heartily to be wished that the German surgeon should at last free himself from such quack-proceedings, for which the patients thank him but little, as the daily-repeated expression of the matter is far more painful and offensive than a suitable incision; I have often heard the patient say, "the cutting is nothing, but the squeezing is unbearable."

When an obstinate discharge is treated by dilatation or by fresh incisions where, through improper treatment, the suppuration is become profuse and of bad quality, the quantity becomes instantly less, the quality improves, and the general state rises in tone. The fearful Sinking of matter can alone be prevented by the means I have mentioned. I have had opportunity to observe during the campaign that there are two forms of "sinking" of matter of which one alone merits this name though both are so entitled. The true "sinking" of matter [Burrowing of matter] consists in the pus formed in one spot sinking according to gravity in the cellular tissue, usually under the fascia—partly infiltrating the cellular tissue,—partly forcing it aside and thus forming large cavities. The second kind consists of new inflammatory processes wherein the part swells, hardens but later fluctuates, if this be opened early no pus is visible but serum or a gelatinous mass, yet these cavities filled with serum stand in connection with the chief focus of suppuration so that occasionally the finger may be passed from the serous into the suppurating cavity. It seems that in these cases the pus-serum had infiltrated under the fascia, causing an acute inflammatory process excited through the acridity of the secretion, while in the true sinking of matter the pus extends itself without causing inflammatory swellings in its path. Only after some days, and after the application of poultices, the incision in the serum-filled cavity commences to yield pus, and to serve as issue for the chief collection. In practice, the difference of the two kinds of burrowing of pus is a matter of importance,—the true one requires immediately a sufficient counter-opening; in the inflammatory form, however, leeches are sometimes able to prevent the further formation of purulent foci.

The Diet of such wounded must of course correspond to the antiphlogistic treatment, consisting of cooling drinks, water-soup [a kind of thin gruel], and a little bread, perhaps for many weeks. Possibly it is superfluous to mention this, but I have unfortunately found that many medical men do not trouble themselves in the least about what their patients eat and drink, or consider it of no importance. I have seen numberless times that the patients received roast or broiled meat, as soon as they had the least appetite again. Frequently the patients think that they must have something nourishing throughout to support their strength, they eat therefore what is given them without having appetite for it. It was well worthy of remark, that this unsuitable diet seldom caused indigestion, but it certainly contributed to keeping up the suppuration and inflammatory accidents, as in many patients a diminution of the suppuration at once followed, after I had put them upon vegetable diet. I recommend therefore in the most urgent manner a greater severity in dieting, and so much the more as later — the dangers of the first week being happily passed by — a more nutritious diet affords so much the better service.

### Further Progress of Gun-shot Wounds.

In favorable cases the inflammatory accidents are not severe, or become rapidly milder. Suppuration lessens, sequestra, which through sticking pain in the wound, and by slight bleeding of the granulations, are known to be entirely free, are sought for and removed,—with the matter the fragments of clothes which had been driven before the bullet are discharged, and are carefully collected by the patient, who thus reckons what may remain. The last piece having come to sight, the suppuration lessens decidedly, yet the wound still remains open, as tertiary sequestra have to be separated in many cases. Many surgeons, if they feel necrosed bone with the probe, now commit the fault of forcibly attempting to keep the wound open, until the sequestra are loose. Fortunately sponge-tents were not at hand, but in their place I found many plugs of considerable length forced into the wound. Their action is entirely prejudicial, they hinder the escape of matter, keep up the inflammation, cause unnecessary pain and generally fail in their object, as the dilatation of the wound by the knife for the extraction of large fragments, still cannot be dispensed with, and small ones escape of their own accord. I have frequently seen that the track of the wound closed shortly after the removal of the tent, although necrosed bone could be felt, — this is explained by the necrosis affecting but a very thin layer, which escapes in minute particles.

Among the dangers threatening the patient labouring under gun-shot injuries of the bones, pyæmia stands foremost. It occurs especially—1st, in such wounded who must be transported far; 2nd, in those who with fresh wounds have been admitted into hospitals, which had been filled with wounded from previous battles, and still contain patients with suppurating wounds. It also seemed that more pyæmic cases occurred in hospitals, which, besides having been employed for cases of other internal disease, had also received typhus patients. 3rd, Surgical operations, in which the medullary canal of the bone is exposed to the influences of air and matter, seem to favour the disposition to pyæmia. Hence it has been very frequent after resection of bones in their continuity in 1848, and after amputation as well in the present, as in the past year. In many cases this began, as it were in an acute form, for it appeared on the fifth or sixth day by shivering fits, jaundice, retching and bilious vomiting and soon ended fatally. Generally, however, it occurred first after many weeks; making itself known, not immediately by shivering fits, but by a peculiar languidness which especially express itself in the glance, by a small quick pulse, and in most cases by want of appetite. The course was very different, in some very rapid in others lingering, so that they struggled with death for months. In the majority of cases death was caused by abscesses in the lungs, yet in injuries of the lower extremities obstinate diarrhœa was often present and frequently ended fatally, the mucous membrane of the large intestine after death being found lined with a diphtheritic layer. Some patients of the last class were saved by the use of acetate of lead with opium. In injuries of the upper extremities, abscesses in the lungs formed rapidly, and indeed with preference to the side of the injury. The shivering fits frequently remained absent a whole week, and were frequently delayed by the use of bark, without the fatal termination being thereby avoided. In some patients who died of pyæmia, there was no rigor, but in its place another convulsive affection, viz. an intractable hiccup, continued, or at short intervals.

The question of parenchymatous hæmorrhage and its cause has already been discussed. A single amputation practised on this account proved fatal rapidly; the ligature of the chief artery had but temporary success: the extraction of extensive loosened sequestra effected sometimes temporary stoppage of the hæmorrhage, but death followed by pyæmia. An arterial hæmorrhage from the wounded brachial occurred in the third week; the humerus having been fractured, amputation was performed; however, the patient shortly died, too much blood having been lost in the first instance, an incompetent medical man who was present not having presence of mind to employ compression.

We have only had six cases of Tetanus among 2000 wounded, only one of whom had injury of bone. Three of these recovered; they had, however, the chronic form of the disease. Baths and moderate doses of opium afforded here excellent service.

---

## GUN-SHOT WOUNDS OF SINGLE BONES.

### Injuries of the Facial Bones.

It is remarkable what slight accidents follow these injuries. I have seen many cases where a bullet had traversed the root of the nose, or somewhat deeper beneath it, without the slightest symptom of concussion of the brain showing itself, as is otherwise frequent in fractures of the nasal bones; these wounds healed rapidly, and without extensive exfoliation. The wounds are just as trifling where the bullet entering the upper jaw of one side traverses it and the nasal fossa, and escapes through the opposite cheek, both antra of Highmore being frequently opened, Still more frequent are cases where the bullet strikes upon and injures the upper maxillary bone anteriorly, and makes its exit in the neighbourhood of the ear, in these cases a paralysis of the facial nerve is common on the injured side, fully disappearing after some months, so that we must probably refer it to a contusion of the nerve.

In one case, the bullet had entered the antrum Highmorianum anteriorly, proceeding behind the ascending branch of the inferior maxillary bone, had in passing out torn away the anterior boundary of the cartilaginous meatus externus and had injured the internal maxillary, there occurred on the night of the eighth day, after removal from Colding to Christiansfelde, a severe hæmorrhage from the ear and from the entrance-wound. The carotid was taken up between one and two o'clock, and the patient recovered. Here also there had been at first complete palsy of the facial nerve, which gradually diminished. In one case the bullet entering on one side and tearing open the cheek, had broken the jaw in pieces, and forced its way into the nasal cavity,—here I had the outer wound united by interrupted suture, after it had cleansed itself;—the fragments found exit through the nasal fossa, after the cheek was fully healed. An extensive loss of substance from the upper jaw-bone including five teeth, healed without leaving communication between the nose and mouth.

Balls remaining in the antrum Highmorianum often give rise to no accidents, as the wound heals rapidly over them. In one instance the flattened bullet came to the surface under the mucous membrane of the fossa canina three months later and was removed by incision.

### Gun-shot Injuries of the Vertebral Column.

Injuries of the spinal processes frequently occurred without serious consequences—without accidents from concussion of the spinal marrow. A case where the posterior portions of the fifth and sixth cervical vertebræ were torn away by a bullet, ended fatally on the fifth day; after to paralysis of the lower extremities occurring on the first day, complete paralysis of the arms had become superadded. In two cases the cervical vertebræ were contused by a bullet, which had entered on the outer side of the sterno-mastoid, and had also bruised the brachial plexus; the paralysis of the arm of the corresponding side was at first so complete that I considered the brachial plexus must have been torn by the bullet, but gradually sensation and motion returned almost fully. In one of the cases the phrenic nerve must have been contused, for during eight days there was great dyspnœa present, and the patient was obliged to remain in the sitting posture; it was at first supposed that the lung was injured, but there were no physical changes on the corresponding side of the chest. In a case of contusion of the cervical vertebræ by a similar shot, there has remained till this moment—a period of four months—stiffness and pain of the neck on motion. In all these cases small sequestra escaped. In a case, where a bullet entering laterally, had severely bruised the third and fourth cervical vertebræ and had not been extracted, death followed by the advance of inflammation in the spinal chord and brain; there was at first palsy of the arm of the injured side, followed by incomplete paralysis of all the limbs and ending in stupor;—antiphlogistic treatment had been entirely neglected in the first instance.

The first of August, I extracted a bullet, which had entered on the sixth of July, between the arches of the third and fourth lumbar vertebræ, and there had become fixed. At first there were no severe symptoms, suddenly there occurred violent pains with cramp in the extremities, having similarity to tetanus and accompanied by delirium. Although not expecting a favorable result I undertook the extraction of the bullet, which was easily performed by the help of an elevator, after dilatation of the outer wound. On the removal of the bullet, the finger could be introduced into the spinal canal; the patient sunk rapidly, and the autopsy shewed inflammation of the spinal chord.

Injuries of the facial bones are only very dangerous, if they traverse the same and strike the base of the skull or the cervical vertebræ These injuries will be treated of, with the Injuries of the Head.

### Gun-shot Injuries of the Lower Jaw.

These injuries occur very frequently, and in the most frightful form. In many cases the bullet had traversed each lateral half of the lower jaw, and on each side had caused comminuted fracture. In one case, all the teeth excepting a pair of molar teeth, were struck out not only from the inferior—but from the superior maxillary bones. In another case a grenade shot caused the destruction of those parts of the upper and lower jaws supporting the incisor and canine teeth. In most cases the bullet had broken the jaw-bone on the one side, and passed out again through the opposite cheek. An instance occurred of simple fracture of the lower jaw, anterior to the insertion of the masseter, the bullet not having entered: I first found this about eight days after the injury, it had been overlooked from the motion of the jaws being unaffected and no dislocation being present.

The treatment of such cases was extremely simple. I merely removed those teeth and fragments of bone, which were either fully loose or were hanging by thin shreds; the discharge of the rest was expected from the suppurative process and they were only removed when this could be performed with ease. In no case was any resection undertaken of the injured jaw,nor any bandage applied, the chief object was cleansing the mouth first by injection of cold—later of lukewarm water. This is positively essential, as the swallowing of matter leads not only to gastric irritation but also to typhous symptoms: as I have twice seen, where, however, through careful cleansing of the mouth with solution of chloride of lime an improvement was soon perceptible.

The results of this simple manner of treatment was very favorable, as not only the majority of the patients recovered, but also the resulting deformities were but slight. The fragments united by bone, excepting in two cases in which there occurred ligamentous union; this circumstance I consider due to the fact that I left the discharge of the greater fragments of bone to the suppurative process and thus the periosteum remaining behind could form new bone. In the case, where the middle portion of the upper and lower maxillary bones were carried away by a grenade shot with considerable loss of substance of the under lip, no union in the middle line occurred either of the soft or of the hard parts. In this case, after a delay of four months, the soft parts which had united to the two halves of the lower jaw and were much retracted, were loosened by me from the bone freely, and after the edges were pared were united in the middle line by the twisted suture. The parts healed, and the saliva could now be retained, however, the total failure of the under-jaw was a special disfiguration.

Three cases of injuries of the lower jaw ended fatally; in the first, a Saxon soldier, the bullet entering near the chin had driven forwards the fragments of the shattered lower jaw under the tongue against the orifice of the larynx. When I first saw the patient he was in imminent danger of suffocation, from which I freed him by removing many sharp fragments from the neighbourhood of the glottis, by means of the left index finger introduced into the mouth: however, sudden death occurred on the fourth day from hæmorrhage:—the autopsy showed that this had taken place from the internal jugular vein, which had been pierced by a sharp fragment of bone. In this case warm poultices had been immediately employed, and very improperly so: on representing this to the acting surgeon, he thought that secondary hæmorhage would have followed under any circumstances. However, the instance to be mentioned proves that this is by no means inevitable; on the taking of Colding a Schleswig-Holstein soldier was shot in the right side of the lower jaw before the insertion of the masseter, the bullet found its way under the tongue to the left side of the lower maxillary bone, to behind the left sterno-mastoid, level with the top of the larynx, where it was perceptible at the first. It had not been attempted to remove the bullet at once by incision, later it was not to be felt. The case proceeded favorably at the commencement although the patient was very much depressed: however, in the third week pleurisy occurred by metastasis on the right side, and proved fatal. On the autopsy an abscess was found behind the left sterno-mastoid, in which lay the flattened bullet near to the vertebral column;—the internal jugular vein was torn by the bullet for a length of five lines on its anterior and outer side from above downwards,—this rent was however, fully closed, as the coats of the vein had applied themselves to it behind and were united, by this means the volume of the vein was become of but half its size. Had the bullet been removed by incision and in the first instance, the danger of hæmorrhage, or of the entrance of pus into the vein would certainly have been very great. The third fatal case was a Dane, whom a bullet had struck at Fridericia, at the left angle of the lower jaw,—it had entered and on account of the swelling could not be recognised; it could be felt from the cavity of the mouth, that the jaw above its angle was broken in many pieces. On the fourth day, a violent arterial hæmorrhage required instant attention, although it could be arrested for the minute by plugging from the cavity of the mouth, the carotid artery was therefore ligatured. Four days later the bleeding recommenced, but this time of venous, and none of arterial character, on extracting now for the first time some sequestra and then the bullet by the mouth, the loss of blood ceased for some days, then returned accompanied by shivering fits and ended fatally. On examination, besides the appearances of pyæmia, there was splintering of the bone nearly as far as the joint. There was, as frequently, no evidence of the source of hæmorrhage. Possibly instead of ligature of the artery, exarticulation or resection might have been preferable, but as the frequent recurrence of hæmorrhage stood in connection with pyæmia, this operation would scarcely have saved the patient.

### Gun-shot Wounds of the Pelvis.

These injuries were always very dangerous, excepting those where the crest of the ilium was struck and shattered. These cases almost always ended favorably—the inflammation being moderate, so also the subsequent suppuration; the sequestra were removed gradually after suppuration had fully commenced and only the discharge of tertiary sequestra in some degree hindered the cure. As a rule in these cases the bullet had not penetrated deeply, or it was early removed. Indeed one case proved favorable, where the bullet had comminuted the anterior superior and inferior spinæ of the os ilium, and had lost itself in the neighbourhood of the horizontal ramus of the pubes, where it yet remains. However, all those cases ended fatally, where the bullet penetrating the pelvis posteriorly, through the thick muscles of the back, had either broken off large portions of the ilium, or had simply penetrated it. The patients after suffering great pain, died with symptoms of pyæmia, after the wound had become sloughy. On examination, besides evidences of pyæmia, the injured bones were found laid bare of periosteum to a great extent around, and bathed by large quantities of bloody, serous exudation. In one case, I was able to extract the bullet from its position in the middle of the ilium, by a tirefond, however, death was not prevented. Deep incisions for the dilatation of the wounds, in cases where the bullet could not be found, were of little avail, as in spite of them, exudation took place beyond the bullet, and it was not possible to effect the discharge of this fluid. Comminuted fractures of the ischium were equally dangerous, where the bullet had penetrated the thigh and taken its direction against this bone; as a rule these fractures were overlooked and only recognised after death, or on a dilatation of the wound rendered necessary by the decomposing sanious discharge. I have seen one case of injury of the ascending branch of the ischium, end favorably after extraction of large tertiary sequestra. The contusions of the ischia also had very bad consequences, and gave rise to obstinate suppuration and hectic fever.

I consider the antiphlogistic—as the only proper—treatment of these deep injuries of the pelvis. As in these cases other foreign bodies are frequently present, as well as the bullet—such as pieces of clothes,—which must necessarily excite suppuration, we must not expect to prevent that process, but it can be so checked by strict antiphlogistic treatment, that an extended formation of sanious pus shall not take place and that the foreign body shall be gradually expelled. However, in these cases the antiphlogistic treatment is generally neglected. pain is relieved by opium and mercury given should pyæmia arise. For the rest, the preservation of life in deep penetrating wounds of the pelvis, is not always to be reckoned good fortune, as a case proved from the campaign of the former year, where the bullet entered in the middle of the right gluteal region, and where after the lapse of a year, no cure had followed; the patient still leads an ailing and painful life.

## Gun-shot Injuries of the Clavicle.

These injuries are not so dangerous, as might have been expected on account of the neighbourhood of important organs. I have met with them in all forms from simple fracture, where the bullet had glanced off, to complete comminution of large portions of the continuity. In one case both the acromial end of the clavicle and the spine of the scapula were fractured. In the case of a lad in Freiburg, who had received a musket-shot from before in close proximity, the acromial end was fractured and a piece of integument of the size of the hand was torn away, this case also was successful; the remainder of the clavicle, which had been drawn upwards half an inch, and had projected very much was again approximated to the shoulder during the process of cicatrization, and the cure resulted without deformity.

The treatment of these cases was extremely simple,—the arm was fastened to the trunk, cold applications at first employed, and oiled charpie after the full commencement of suppuration; the fragments of bone were only withdrawn when become fully loose. Even after a loss of two inches in length from the continuity of the clavicle, bony union took place, and the subsequent difference in length of the two clavicles was scarcely remarkable.

Two cases proved fatal, these were attended by peculiar circumstances. The clavicle was obliquely fractured in the centre and the outer fragment was driven into the soft parts backwards and upwards, reaching with its ragged end the brachial plexus. Replacement could not be effected. The patient had extreme pain in both arms, neither of which was he capable of moving, with exception of slight motion of the fingers. I saw him first on the eighth day, when he had already had rigors and was in high fever. On dilating the wound it was possible to withdraw the outer fragment from the soft parts, and to saw off its broken surface,—this was also performed upon the other fragment; the two pieces removed by the saw fitted one another, so that there had been no comminution of the bone,—the bone thus lost was one inch in length. Death occurred after some days—on examination, contusion was found of the brachial plexus, however, the spinal marrow was not inflamed, as I had supposed from the pain and loss of motion in both arms. Had the operation been performed in the first instance, a good result would have doubtless followed; reposition should certainly have been effected, had dilatation even been necessary for that purpose.

Injuries of the clavicle seem to me specially suitable for proving, that expectative treatment—without surgical interference at first—is by far the most preferable, and that the early extraction of sequestra, or resection is not necessary unless under peculiar circumstances.

### Gun-shot Injuries of the Scapula.

The injuries of the shoulder-blade arise not only from bullets piercing from the side or from behind, but also very frequently from such as penetrating the pectoralis major from before, reach the axilla and strike the bone from within outwards. In one instance I observed an injury of the inferior angle of the scapula from a bullet, which entering on the inner side of the arm close above the elbow, had coursed along the vessels and again became free between the scapula and vertebral column.

Fractures of the scapula are only very dangerous, when the fissures extend into the shoulder-joint. In such cases, there does not follow as might be expected, an acute inflammation of the joint, for it does not swell and is not painful on pressure, but the wound takes on a bad character, and pyæmia results. The diagnosis of the fissures into the joint being difficult, even when it is known that the scapula is broken, all these cases must be treated antiphlogistically, and that very strictly. Dilatation should not be employed, as it allows freer access of air to the injured spot,—should it be required for the extraction of foreign bodies, it should be first employed, when it has appeared in the course of the case that the joint is uninjured. Burrowing of matter requires particular attention in injuries of lower portions of the scapula,—they must be put an end to by making a transverse incision of many inches in length below the shoulder-blade through the latissimus dorsi, thus it becomes unnecessary to lay open the sinuses singly. One case proved successful, where the bullet entering through the pectoralis major, had pierced the scapula and lodged in the infraspinatus muscle; its extraction, followed by that of osseous fragments, was not effected till after the formation of pus in the fossa infraspinata. Immediate hæmoptysis, pleuritic effusion and great dyspnœa had evidenced the injury of the corresponding side of the chest, but the extent of injury could not be further determined. In another case, the bullet entering the fossa infraspinata from behind traversed the scapula and lodged in the fossa subscapularis, where I could feel it with the finger,—the finger was held fast by sudden muscular spasm, so that it was some instants before I could withdraw it, judging from this circumstance that there must have been extended fissuring, I made no further attempt for extraction of the bullet at the time; but after some days I was able to extract it—much altered in its form. Later pyæmia set in, and on examination fissures were found extending into the joint. The earlier remova.

of the ball by means of the crown of the trepan, would not have tended to preserve life. In all these cases, the nearer the injury of the scapula is to the joint, so much the more likely are fissures to extend into the same.

## Gun-shot Injuries of the Humerus.

These must be divided into the injuries of the shaft and of the articulating ends. Those of the shoulder-joint are not rare, occurring from the head of the humerus being struck by a bullet from before, from outwards or from behind, and also from inwards, the last penetrating near the coracoid process and traversing the axilla; the brachial vessels and nerves usually escape in all these injuries. Bullets grazing the shoulder-joint, often traverse a portion of the deltoid without opening the joint; whether the articulation is opened or not cannot always be determined on examination of the wound, but is evidenced in the course of the case by inflammation arising in the joint. This uncertainty in the diagnosis is of no great importance, as in either case antiphlogistic treatment must be employed, and injury of the capsular ligament calls for no special treatment. With care the cure takes place by anchylosis after long-continued suppuration, but should the last prove dangerous to life, resection of the head of the humerus is to be performed.

Should the bullet have entered from before or behind, it may break the head of the humerus in many pieces, groove the same, or cause an indentation or lodge itself in the substance. The extent of the injury is generally recognisable on the introduction of the finger and on employing movement of the injured bone; the depressions alone, are difficult to discover, if the superincumbent soft parts are not lacerated, or if swelling has already taken place. As these depressions are complicated with fissures into the joint, suppuration of the last happens at a later period. Extensive comminution of the head of the humerus does not necessarily demand its resection, as a cure can occur by anchylosis, after the fragments loosened by suppuration, are discharged; as however, with the help of resection, the mobility of the arm can be preserved, it is preferable to undertake this operation at once, or after suppuration has been fully established.

Balls arriving laterally are more dangerous than those last mentioned. If superficial they may be withdrawn by the clothes carried in before them, as by the shirt, but if not lodging in the head of the humerus it is to be feared that they have proceeded further,—either into the cavity of the chest, or if passing towards the scapula, this is probably shattered and the bullet lodges under the same, or in the fossa supra- or infra-spinata. In these cases the diagnosis is difficult especially with reference to the injury of the scapula, and hence it is better to adopt expectative treatment, and merely employ antiphlogistic measures. For in extensive injury of the scapula, resection of the head of the humerus is not successful, as I have once seen,—death was preceded by parenchymatous hæmorrhage. The delay may be so much the more justified, as resection can be performed with the best results, after suppuration has been fully established.

Those cases are the most difficult, when the shoulder-joint has been injured internally and inferiorly by a bullet entering in the axilla. Every thing here depends upon deciding in the fresh state of the wound as to the extent of injury to the bone. A case occurred, where the medical man in attendance supposed that an injury to the ribs existed, and where on the autopsy it was found that the anterior costa of the scapula was split off including a portion

of the glenoid cavity, and that the head of the humerus was grazed. In this case, hæmorrhage about the twelfth day, lead us to take up the subclavian— however, the bleeding recurred and the patient died. Had the piece of the scapula been removed in the first instance, after the resection of the shoulder-joint, the case would have probably proved favorable.
It is probable that in these injuries, no cases give such extended fissuring of the scapula, as the lateral gun-shot wounds.

Balls striking the shaft of the humerus in the neighbourhood of its head, cause descending fissures—the spongy head of the humerus is not so disposed to split as the brittle shaft. It is only when the bullet hits the junction of the epiphysis and shaft that the fissures proceed both upwards and downwards, hence, a resection of the head of the humerus should not be lightly entered upon. The fissures of the humerus proceeding far downwards would require the removal of a lengthy portion of the continuity, giving rise to burrowing of matter and necrosis, by which the recovery is very much delayed; should therefore amputation be not indicated, the case is rather to be left to nature. I have seen one case end favorably, where the fracture took place an inch and a half below the tuberosities of the humerus. In a case where the simultaneous injury of the soft parts rendered preservation of the limb impracticable, I directed amputation to be performed—instead of exarticulation—half an inch below the tuberosities of the humerus, and this with the best results. I consider Larrey's view, that exarticulation is preferable to amputation, to be erroneous; it could only be correct, if the fissures in the humerus were usually found to extend into the shoulder-joint. But as this is not the case, when the bullet has merely struck the shaft, amputation must be preferred as causing less deformity.

If resection of the humerus is performed in the ordinary way, by an incision between the two tubercles of the head, it is not difficult of execution,— the long tendon of the biceps is readily drawn aside by means of a hook, and the operation then proceeded with; but after this operation the secretion from the wound does not readily escape. I therefore introduced a new method after the battle of Fridericia—through Dr's Francke and Hermann Schwartz— in which the articulation is opened posteriorly by a crescent-shaped incision commencing beneath the acromion, and proceeding backwards and downwards. Thus the secretions escaped with facility as the patient lay, and the success fully justified the expectations. Burrowing of matter did not take place, and the patients overtook in their recovery others who had been operated on months previously. I consider it proper therefore to recommend this innovation urgently to the notice of surgeons. The preservation of the long tendon of the biceps is somewhat more difficult, but it is still practicable.

Injuries of the shaft of the humerus are always of a dangerous character on account of the tendency in this bone to occurrence of long fissures. If the soft parts are much lacerated, it is better to perform amputation; but if they are merely penetrated, the radial pulse perceptible and movement and sensibility not destroyed in the hand—an attempt should be made towards preservation of the limb. The chief risk under these circumstances consists in the severity of the inflammation in the soft parts, along the course of the vessels on the inner side of the biceps; this must be combated by abstraction of blood —local and general—and cold applications; should not amelioration soon result, incisions must be made. A chief difficulty in the treatment is the position and retention of the arm,—if placed on a splint or cushions near the patient, it is affected by every motion of the trunk, and it is therefore better to fasten it to the thorax. For this purpose a chaff cushion covered with oiled cloth is placed between the arm and chest, and retained by a many-tailed roller. The same kind of bandage is used, over a good allowance of soft charpie, to the injured limb; a concave, softened splint, fitted to the arm and suitably defended from moisture, is placed posteriorly and fixed to the chest by a sufficient roller or cloth.

Extensive resections [of the shaft] of the humerus should never be practised in fresh wounds, the attention should be confined to the extraction of fully loose sequestra and to the removal by the saw of sharp points.

Injuries of the humerus in the neighbourhood of the elbow may end most favorably if the fracture is simple, as I have once observed in a case where the bullet caused a simple, tranverse fracture close above the condyles of the humerus, without displacing the fragments. A case of comminuted fracture of the bone for two inches and a half in extent, where resection was performed, proved fatal; while an entirely similar case, where the expectative treatment was adopted, ended favorably but with anchylosis of the elbow-joint.

### Gun-shot Injuries of the Elbow-joint.

These injuries occur very frequently,—in general the bullet had struck the ulna not far from the olecranon, the man being in the act of shooting or loading. Occasionally entering the soft parts of the fore-arm and pursuing its course for some way through the muscles it had, as a grazing shot, shattered the radius or ulna. In one case the bullet had passed transversely beneath the flexors, breaking the coronoid process in its course and at the same time tearing the radial and ulnar arteries; violent arterial hæmorrhage occurred on the fourth day and hence I caused amputation of the arm to be performed, after assuring myself of the injury to bone, by the introduction of the finger. In another case of violent hæmorrhage from a gun-shot wound on the volar side of the forearm close below the elbow, I found the medical man on the point of taking up the brachial artery; I decided from the kind of swelling that injury to bone was present, and by introduction of the finger that the radius was shattered—which had been overlooked previously—the arm was therefore amputated. Otherwise I have met with but three cases—where the fracture opening the elbow-joint extended so far on the humerus, or where the soft parts were simultaneously so much lacerated—that preservation of the arm could not be thought of. All other bullet-injuries of the elbow-joint, twenty-two in number, were of such a kind that amputation appeared uncalled for. The bullet had struck the radius and ulna more frequently than the condyles of the humerus, this must arise from the fact that these injuries generally occur during extension of the arm and semi-flexion of the fore-arm, in which state the condyles of the humerus are covered by the radius and ulna.

The course of the wounds of the elbow-joint,—if not being recognised they were left to nature—was somewhat as follows; the joint became very much swollen, at the same time a tumefaction appeared along the vessels of the arm internal to the biceps, being especially severe if the condyles of the humerus were injured. For the first day or two the wounds remained dry, and then yielded a bloody serum,—on the fifth or sixth day pus appeared, escaping from the wound in less quantity but breaking through the capsule in numerous spots, at the same time forming extensive sinuses in the arm and forearm. In a case where the real nature of the disease was overlooked during five weeks, abscesses had formed in the lungs, pus was expectorated, and the patient was hectic in a high degree. This case proved fatal by the advance of the pulmonary disease.

In all cases where recognising an injury to bone in the elbow-joint, I never hesitated in recommending resection as a rule. The joint was opened posteriorly, with preservation of the ulnar nerve, by making a longitudinal incision in the course of the nerve [on its outer side], upon this fell a second one, perpendicular to the first, entering the joint between the outer condyle and head of the radius, proceeding over the olecranon, yet dividing the triceps tendon,—so that by this second incision the joint was opened. After that, with constant preservation of the ulnar nerve, the ligaments of the joint were fully divided posteriorly, so that it gaped freely, the loose fragments were extracted, and those still attached removed by the knife or by Cooper's scissors. Had the ulna alone been shattered, the corresponding portion of the radius was removed, yet never more than the head of this bone, or perhaps also the neck, even when the ulna had been comminuted. Indeed it never happened that the whole continuity of the ulna was destroyed to a greater extent downwards than the neck of the radius, at least a portion projecting thus far. So that by division of the bone at this spot, transversley by the saw, it was attempted as much as possible to preserve the bones of the forearm of the same length. Had the ulna been divided at the commencement of uninjured bone and a corresponding piece of the radius removed, in many cases three or more inches in length would have been lost, the operation would have become very difficult and severe, and the consequences probably much worse, as neither a new joint nor anchylosis would have resulted, if indeed the patient had escaped with preservation of his life or his limb. I have never had occasion to repent this sparing method of treatment, but, on the contrary, have assured myself that new formation of bone occurs from the preserved portion of the ulna, whereby it reacquires its usual volume.

If the radius was comminuted below the neck, the fragments were extracted and the uneven projections removed by the saw; the ulna was not sawn off at the same level, but merely a piece removed from the olecranon, as otherwise this protruded in a troublesome manner during and after the healing. As on the ulna depends especially the strength of the elbow-joint, it is not necessary to saw this off so deeply, as the radius.

After the ulna and radius had been freed of their fragments and rendered even by the saw, a portion alone of the trochlea humeri was sawn off, if but little of the ulna had been removed. In many cases the humerus was untouched or merely the cartilage shaved off by the knife,—this last procedure seemed to have no influence towards a more rapid recovery, as Zeis has expressed his opinion that it might. Had the injury of the humerus involved the elbow-joint, similar methods were adopted, as with the radius and ulna. Generally one or other condyle was shattered and the fissuring proceeded above the capsule,—resection at the cessation of the fissures would have necessitated a loss of substance of from two to two and a half inches. The humerus was therefore sawn off, after the removal of fragments, at the place where the fissures had nearly terminated, and by this means an inch in its length was frequently preserved. In these cases, if the bones of the forearm were uninjured, a piece was still removed from the olecranon, but otherwise no interference allowed. A portion of the wound was united by interrupted suture, the lower portion being left open for the escape of secretion.

The result of resections of the elbow-joint were extraordinarily favorable, while very many of the cases of amputation were lost, those of resection of the elbow with few exceptions, recovered and preserved the motion of the hand. In the majority however, incomplete anchylosis took place.

Conducted in this manner the resection of the elbow-joint is a treatment just the same as that long considered proper in all compound fractures; that is, the withdrawal of fragments and the removal by the saw of projecting points. It does not seem advisable to me, to cause greater losses of substance purposely, with a view of more certainly preserving the motion of the elbow-joint; partly —because experience teaches, that anchylosis of the arm in an obtuse angle does not interfere much with its use, partly— because the operation requires so much more extensive separation of the muscle, the farther from the joint that the bone is divided. Besides this I am not convinced that the preservation of motion in the elbow-joint depends directly upon the extent of bones resected, as motion was preserved in some cases where less had been taken away than in others in which anchylosis took place. Probably the movement to which the resected joints are subject to, earlier or later, have a great influence in this respect; the patients not remaining under my view, I was unable to take the direction of this part of the treatment.

The oftener that I myself performed resection at the elbow-joint, or had it performed, so much the less could I decide to leave compound fractures of this joint by gun-shot, to nature, as Guthrie seems sometimes to have done. The fragments are of that kind usually, that their extraction seems almost impossible, even after suppuration has been fully established, without freely laying open the joint. This especially holds good of fragments of the ulna;—thus, it occurs frequently that the processus coronoides remains in connection with a long splinter of the ulna, supporting itself above, on the trochlea humeri—below, on the now long, oblique, broken surface of the ulna; —in this case it is often difficult to extract the sequestrum, even after opening the joint.

In the after-treatment of this resection I consider it of the greatest importance that the arm constantly remains at rest, upon a flat, padded, wooden splint, protected by oiled-cloth, and reaching from the upper third of the arm to the points of the fingers; forming at the elbow an obtuse angle, and having in the portion corresponding to the inner condyle, an aperture the size of a crown to obviate pressure upon the part. The wound must be dressed without moving the arm. The cure may often occupy three months, yet it succeeds frequently much more rapidly. As regards the result, it is of no consequence whether the resection, has been performed in the first forty-eight hours, or after the full development of suppuration.

## Injuries of the Ulna and of the Radius in their Continuity.

These cases occurred extraordinarily often and resulted favorably almost without exception; even including a case where the bullet had crushed both radius and ulna, having passed transversely from side to side through both bones. Only one case required amputation of the forearm, here the lower end of the radius was shattered,—hæmorrhage from the radial artery, wounded simultaneously, had weakened the patient, whose state was rendered still more critical, by the inflammation and suppuration being very much aggravated by unsuitable employment of compression; had the vessel been taken up at the proper time the hand would have been saved.

In a case where the ulna had been struck, three inches below the olecranon, fissures seem to have extended into the joint, for it suppurated; it was cured but with anchylosis, after an incision had been practised near the head of the radius, so as to arrive at the articulation. In another instance where the ulna was struck in the middle, extended fissuring took place according to the long axis of the bone. The inflammatory complications were severe and obstinate, were however, fortunately held in check by leeches, and union occurred after the discharge of long, pointed sequestra.

The treatment under these circumstances was extremely simple. The fore-arm was placed upon a flat splint from the elbow as far as the finger ends, all compressing dressings were carefully avoided, at first cold, later warm fomentations were employed. On decrease of the suppuration, the application of a flannel roller had excellent effect. The sequestra were withdrawn after becoming fully loosened by suppuration.

### Gun-shot Injuries of the Wrist.

The bullet has here shattered the radius and ulna and opened the wrist-joint, or has pierced the first range of carpal bones. Not unfrequently both parts are injured by the straight or oblique passage of the bullet through the joint. The chief danger under these circumstances consists in the subsequent violent inflammation of the wrist-joint and of the carpus and its joints; this affects the system in an especial manner and peculiarly disposes it to the occurrence of pyæmia. This evidently depends upon the articulation being surrounded by many tense tendons, which being retained in position by the annular ligament, afford a considerable resistance to the inflammatory swelling,—hence the intensity of the pain and that of the inflammation is much increased. It is only after the setting in of suppuration that the ligaments of the joint, and the surrounding tendons can yield. The suppuration points in various places, generally between the thumb and index-finger, and on both sides of the joint. In the mean time however, formation of pus has taken place in the cancellated structure of the bones, hence the danger of pyæmia and the fact that even amputation does not always save life.

Yet I am by no means of the opinion that gun-shot wounds of the wrist necessitate amputation, unless extensive laceration is present, which of itself demands this measure. The after-treatment of these wounds was generally defective,—as a rule abstraction of blood was quite neglected,—the surgeon allowing himself to be deceived by the moderate swelling at the commencement—contented himself with applying poultices,—while I am satisfied that by vigorous abstraction of blood, by the application of ice and by opium, the dangerous advances of the inflammation can be arrested and amputation be obviated.

In one case in which the wounded wrist was tense and extremely tender to the touch, the pulse small and rapid, the tongue dry and the patient delirious, I was the means of saving the man's hand and probably his life also, by an incision, which was made on the outer side into the wrist-joint; after this incision all the threatening symptoms disappeared suddenly. In another case, where a similar incision was put in practice after the tense swelling had again become lax, the local symptoms improved, but the patient died from abscesses in the lungs.

### Gun-shot Injuries of the Hand.

Injuries of the second row of carpal bones are far less dangerous than those of the first range, yet by negligent treatment they may have equally unfavorable consequences. My pupils at Freiburg will remember the fortunate cure of a Russian tailor, who having fought in the squadron of Herwegh, received a bullet through the second row of carpal bones and yet recovered without the least interference with the motion of the hand.

Still less dangerous are the injuries of the metacarpal bones, whether one or many are shattered. I have met with a case where a bullet had transversely traversed and broken the metacarpal bones of the four fingers, where nevertheless the hand was preserved and in completely useful condition. The healing occurs with some shortening of the metacarpal bones, the slight discharge of sequestra being very striking; all operative interference should be unthought of.

Gun-shot injuries of the fingers are very frequently met with;—in many cases exarticulation of the injured finger was at once performed with good result,—on the other hand all exarticulation of the fingers which were performed after an interval of forty-eight hours, were followed by very violent inflammation, obstinate suppuration and not unfrequently by stiffness of many fingers or of the whole hand. In accordance with my formerly expressed principles [If the operation cannot be performed before inflammation has set in, the full development of suppuration must be waited for.] I have not performed or encouraged the performance of exarticulation of the fingers in the field, and have thus avoided the inflammatory accidents which are consequent upon operations of this kind when performed too late. I have never seen tetanus arise in these injuries.

In all injuries of the hand a splint must be employed, and poultices are earlier indicated and must be longer continued, than in other parts of the body. Local baths may be very readily applied and with great advantage in these cases, as well as in injuries of the fore-arm,— they may be employed sooner in the hand than in the arm.

### Gun-shot Injuries of the Femur.

Among these the most dangerous are the injuries of the head, the neck and the neighbourhood of the trochanters. In one such case, death followed rapidly with symptoms of pyæmia, in two others— after protracted suffering. In a fourth case Dr H. Schwartz performed resection under my direction,—the bullet had penetrated anterior to the trochanter major and passed obliquely inwards, the sufferings of the patient were excessive and the suppuration so free, that it was probable that the hip-joint was involved. The operation consisted in making even the lower fragment with the saw and in exarticulating the upper one,—it was neither difficult of execution nor attended by hæmorrhage, yet after some days the patient died of pyæmia and on the autopsy it was found that a part of the ischium had been shattered by the bullet.

I could never decide in favour of exarticulation of the hip in these cases, expecting indeed nothing from it, as the result of amputation of the thigh had already proved so unfavorable. I should in any case, as Oppenheim, give the preference to resection, and that after the occurrence of suppuration, because then the operation is much easier, and because after some continuance of the formation of matter it can be better expected that the patient will not sink from pyæmia. This operation is considered to be much more difficult than it really is, because it is usually known as performed on the dead body.

That I did not perform the operation in the remaining three cases, depended on particular circumstances; if left to themselves, all injuries of the femur close to the hip-joint must end fatally on so large a joint suppurating, but the diagnosis of the condition present is not without difficulty. When the bullet has struck in the neighbourhood of the trochanter itself, it is not so easy to decide whether the injuries extend to above the trochanter, or not so far. In all the cases that I have seen, extensive comminution was present, and the joint had even suppurated in those cases where the fissures did not extend to within the capsular ligament. The usual symptoms of fracture of the neck of the femur —the shortening, and the external rotation of the foot—failed in the cases I observed; this is due without doubt to the fragments hanging together better on account of the partial preservation of their fibrous covering. In one of the cases where death followed so rapidly by pyæmia, the patient could perform considerable motions of flexion and extension with the leg, although the neck of the femur was broken; in another case the fragments fitted so well together that the patient did not experience the least pain and the leg could be moved without causing crepitation, so that the surgeon considered that bony union had already taken place, although I drew his attention to the fact, that this could not be possible on account of the profuse discharge. On examination after death, it was found that the seat of fracture was necrosed, and that a portion of the trowsers lay between the fractured ends. In this case the bullet had been removed by an incision from a purulent deposit behind the trochanter major.

### Gun-shot Injuries of the Shaft of the Femur.

These injuries belong to the most dangerous gun-shot wounds of the bones. Nevertheless they do not always demand amputation; this dogma was taught formerly by military surgeons, who gained their experience under circumstances which were very disadvantagous to recovery, where the patients were obliged to be far removed and treated in over-filled hospitals.

In the course of this campaign a considerable number of such wounded have been cured with the preservation of a useful limb. The condition under which this may be hoped for, are especially those where the injury consists in simple division of continuity, without extended shattering, and where the bullet has traversed the limb. Injuries from heavy projectiles generally cause comminution of the femur, yet on examination it is not always easy to be recognised,—the presence of a fracture indeed is readily discovered, but not the extent of fissuring, if swelling has already taken place—as it does here very quickly. On this account in fractures of the thigh by heavy projectiles, little hope is generally felt for the preservation of the limb when a wound is present, or in those cases where the fracture has taken place by a piece of exploded bomb or a grazing cannon-ball, without division of the soft parts. An instance of the last has not occurred to me.

Musket balls which strike the femur may cause comminution at the point struck, but frequently the bone is merely broken and not splintered. The most favorable case is when the femur is struck on its outer side by the bullet, which has entered anteriorly, and has made its exit on the opposite side. This direction of the canal of the projectile is so much the more favorable, because the great vessels are neither contused, nor do they lie within the limits of the inflammatory or suppurative processes; and again, because the escape of the secretion from the wound readily occurs, when the direction of its canal is perpendicular in the recumbent position. If the canal is oblique, the case is so much the less favorable, as there is greater injury to the soft parts, and the suppuration is more extensive. It must be remembered also, that the danger increases the nearer the fracture is to the trunk—partly from the greater mobility of the upper fragment according as it is shorter, partly from the increased risk of inflammation, suppuration and phlebitis.

The danger of these injuries of the thigh especially depends on the subsequent inflammation and suppuration; for instead of a plastic process rapidly occurring in the neighbourhood of the broken ends, as in ordinary simple fractures,—in these cases the injured parts pass far too readily to suppuration. The surface of the fracture becomes necrosed, and thus—until after discharge of the sequestra—they generally want the capability of uniting themselves by callus. Usually suppuration persists and the fracture remains ununited; sometimes, however, the callus extends over the necrosed fragments and consolidation follows with diminution of the suppuration; still the discharge of the sequestra is exceedingly difficult—science helps but little, and nature herself is often for years unsuccessful. It may be well conceived therefore, that army-surgeons have in the majority of these cases considered amputation as necessary; especially under unfavorable circumstances and after a long transport. In our time, when the tendency to pyæmia is so marked, life also is brought into the greatest danger by amputation, and it appears to me that the question is not yet decided, whether under favorable circumstances it is better to attempt preservation of the limb, or to practise amputation in all gun-shot fractures above the middle of the thigh.

It seems to me to be of the first importance not to subject such wounded to a removal to any distance, but to bring them on a litter to the next house, and there treat them, even at the risk of their becoming prisoners. We are indebted to the care of the Danish surgeons for the favorable result of four fractures of the thigh by musket balls; these soldiers being wounded at Fridericia, and treated in the town itself, all of them have full use of their limb, and in two of them no shortening has occurred. Similarly we have returned to the Danes two patients recovered from such injuries. The disadvantages of removal in wagons over rough ways cannot be overcome by any kind of dressing in such fractures,—art and humanity demand therefore, that in the wars of civilised nations such patients should be left in the neighbourhood of the battle-field, and it is not to be doubted that the universally honourable character of surgeons affords sufficient ground for the belief that those falling into the enemies' hand will find as careful treatment as from their own medical men.

Should preservation of the limb be determined on, the suitable position of the patient is next to be considered. When no swelling has come on, the proper direction should be given to the limb and if necessary by the help of chloroform. Two long chaff-cushions with two suitable splints are to be laid on either side of the limb and fastened by a bandage. To retain the foot, the leg rests on a board with a foot-piece to which the foot is fastened, or the foot is to be supported by suitable cushions. A roller around the thigh is not advisable, as it hinders ready access to the wound. Introduction of the finger and long search after foreign bodies should be altogether avoided, the dressings of the wound should be applied with the greatest care, without pressure or raising the limb, indeed without entire removal of the splints, of which one at least should always remain in position. The strictest antiphlogistic treatment is necessary. The inflammatory period being favorably passed, the number of splints may be increased, with the application of a many-headed roller. In case the extended position of the limb is insupportable, the double-inclined plane must be employed.

If profuse or obstinate suppuration occurs, the only means of saving the patient, is by giving up all attempts to preserve the normal length of the limb, and by leaving it to the contractile force of the muscles, which then brings the necrosed ends to override one-another, so that non-necrosed portions of the bones come into contact, and as I have often seen, quickly unite with rapid lessening of the suppuration ;—within eight days I observed under such circumstances, a decided improvement in the condition of a patient. For this object, the leg is to be brought into a semi-flexed position, care alone being taken that the point of the foot does not change its place, this is easily attained by a few cushions.

### Gun-shot Injuries of the Knee-joint.

Should a bullet have entered the knee-joint, it is in general easily recognised as soon as the finger is introduced, by the extensive splintering which it has caused. These cases with splintering require immediate amputation, yet are not always easy to diagnose, for the bullet may have impinged upon either the condyles of the femur, or on the tibia—have caused merely a depression or none whatever—and then have glanced off in another direction. In such cases the finger does not always discover the bone exposed, and yet fissures may extend into the joint from the point struck. The further course of a bullet thus glancing off is frequently very peculiar,—I have often removed bullets by incision from the popliteal space, which striking near the patella, had coursed under the integument for a third of the circumference of the limb. It is equally difficult to determine the injury when a grazing bullet has opened the capsular ligament; as if the opening is small the escape of the synovia is by no means speedy.

The slowness with which inflammation and suppuration occur in the joints under antiphlogistic treatment is well worthy of remark,—I have frequently known many weeks elapse previously to the appearance of the symptoms. At last the presence of a foreign body in the joint excites suppuration, and amputation becomes inevitable. As experience has taught that those wounds of the joint in which it had been extensively opened, have been the most inclined to terminate favorably, amputation not having been practised, I was led, on account of the frequent fatal termination of amputation of the thigh, to make the following attempt:—in the case of a young man, a musket-ball had struck the joint near the patella externally, and made its exit three inches distant backwards and upwards, comminuting a portion of the outer condyle, or rather deeply grooving it. At first it was doubtful whether the joint was opened, when this became certain I laid open the track of the bullet, removed many small fragments and made on each side of the joint an incision of two inches in length through the soft parts and the lateral ligaments. From these openings the puriform synovia readily escaped, and the condition of the patient was for some weeks everything that could be wished. The suppuration decreased in a short time and the wounds had a healthy appearance, but the patient died of abscesses in the lungs.

I have not undertaken Resection of the knee-joint, because it affords little hope even under favorable circumstances, and because in the majority of cases it cannot be certainly known how much of the bone should be sawn off.

### Gun-shot Injuries of the Leg.

A bullet may pass through the middle of the upper part of the tibia without causing comminution;—of this injury I have seen two examples resulting favorably, although the suppuration of the soft parts was very extensive and required many incisions. If the bullet remains fixed in the bone, it must be removed by a tirefond, and when necessary by enlarging the canal by a sharp gouge.

If both bones of the leg are struck by a ball, the splintering as a rule, is so great, that amputation must be performed. One such case however terminated successfully, where the wounded man had not been transported far

Those cases also give no good prospect of saving the limb, in which the tibia being struck, on the anterior surface of its middle or lower portion, the bullet has continued its course through the bone and made its exit posteriorly through the soft parts. The fissuring of the bones in these cases generally extends much further than expected, and thence profuse suppuration takes place accompanied with great risk of pyæmia. Resection in the continuity of the tibia with removal of the fragments, has neither been successful in the former nor in this year's campaign. At the same time expectative treatment generally resulted in pyæmia.

I should therefore be inclined to recommend early amputation in extensive splintering of the tibia, even should the fibula be whole, had not amputation itself so often led to pyæmia; however as regards the last point our experience of this year cannot be depended on, as we were seldom able to amputate within the first twenty-four hours on account of the necessity of transporting the wounded so far. Where such early amputation can be practised, its result must be much more favorable than in expectative treatment.

Gun-shot injuries of the fibula alone in its continuity end well and have no great attendant risk, if the discharge of the sequestra is patiently left to the suppurative process; special dressings are not requisite. Otherwise, as regards injuries of the leg through bullets, the same remarks hold good as for ordinary fractures. For fractures of both bones, the apparatus of Heister is the most convenient. If the fibula is uninjured, the limb is to be laid on its outer side on a padded pasteboard splint.

Injuries of either of the Malleoli by a grazing shot do not by any means demand amputation, but heal as I have often seen with incomplete anchylosis of the ankle-joint, if however, the tibia itself is struck and shattered, amputation must be performed.

### Gun-shot Injuries of the Foot.

The most dangerous are those caused by fragments of bombs, I have seen the whole number prove fatal, either by mortification and rapid collapse or by tetanus.

Injuries of the foot by musket-bullets are not so dangerous, and heal extraordinarily well under favorable circumstances. In Freiburg I saw a case result favorably, where the bullet entering the heel made its exit on the dorsal surface of the foot, it had completely crushed the calcaneum and os cuboides in its course. By means of the application of ice continued nearly fourteen days, I held the inflammation in check, and removed gradually, after suppuration had begun, the osseous fragments which together presented almost the entire mass of both bones. These restored themselves so fully, that the form of the foot was completely preserved,—no anchylosis of the ankle-joint took place, although it had been opened and had discharged synovia. As we had no ice during the campaign I did not expect to be able to reckon on similar results. In a case where the astragalus and calcaneum were shattered, I performed Syme's operation with success. The same operation was employed in a similar case where the calcaneum was shattered and the bullet remained fixed in the os scaphoides,—the case ended by pyæmia. In the first case I had only removed the malleoli, in the second the cartilaginous surface of the tibia was simultaneously sawn off, as Syme has described the method; possibly this circumstance had favored the occurrence of pyæmia, as it took origin from the bone.

Grazing shots with exposure and injury of the calcaneum heal successfully under simple treatment. Injuries of the tarsus by musket-balls are also not very dangerous and heal after extraction of the bullet with unexpected facility and without perceptible exfoliation. Dr Franke lately discovered and extracted from the tarsus, a bullet which had remained there a whole year, and although nearly included by the surrounding parts, had kept up only a trifling fistula.

Of the many gun-shot injuries of the Metatarsus which I have seen, one only proved fatal. The patient was consumptive, and in the first instance was not very aptly treated, as long search was made for the bullet, which had not even pierced the boot. Indeed an incision had been made in the sole of the foot, on the head of the second metatarsal bone which was mistaken for the bullet. I have seen a similar mistake committed at the knee, where the head of the fibula was cut down upon as the bullet, although two openings existed, —one of these having escaped notice. All remaining cases of injuries of one or more metatarsal bones ended well, without any operative interference.

As regards injuries of the Toes, the same treatment holds good for them as for the injuries of the fingers,—amputation is unnecessary as recovery readily occurs without its employment, as however, much pain is occasionally experienced, the use of opium must not be avoided and poultices should be employed in good time.

# ON RESECTION IN GUN-SHOT INJURIES.

OBSERVATIONS AND EXPERIENCE

IN

SCHLESWIG-HOLSTEIN.

# ON

# RESECTION IN GUN-SHOT INJURIES.

## OBSERVATIONS AND EXPERIENCE

IN THE

## SCHLESWIG-HOLSTEIN CAMPAIGNS OF 1848 TO 1851.

BY

DR. FRIEDRICH ESMARCH.

PRIVATE TUTOR IN THE UNIVERSITY OF KIEL, FORMERLY SURGEON IN THE SCHLESWIG-HOLSTEIN ARMY.

*(Slightly abridged, and some of the Illustrative Cases are not translated, but all will be found in the Tables).*

KIEL :
CARL SCHRODER & CO.
1851.

# PREFACE.

Hardly any theatre of war, during the last few years, can have offered a better opportunity for surgical observations and experience than that in Schleswig-Holstein. Two of the most celebrated German surgeons were successively surgeons-in-chief, and, from their direct instruction, and by good use of the valuable opportunities at hand, a number of younger medical men have since proved themselves to be excellent surgeons. Their discoveries in military surgery belong to its history and commence a new era in the same. It requires no excuse, therefore, for publishing our experience in the operation of Resection,—upwards of three hundred surgeons, from every part of Germany, have witnessed the favorable results, especially as regards resection of the elbow-joint.

The prejudices of former surgeons against resection of joints—considering the operation to be but seldom applicable,—can no longer be accounted of force, since our experience has proved that this operation gave favorable results in circumstances where amputation in general proved very unsatisfactory. Every military surgeon must consider it his duty, in future, to practise resections of joints as frequently as amputations have been performed up to the present time.

The non-necessity and danger of primary resections in the continuity is not less manifest to us from three years' comparative observations than is the positive utility of resections of joints; resection in the continuity should be confined to the narrowest limits.

The Author must explain why the publication of this work was not left to more able hands. He was encouraged to undertake it by his father-like friend, the Surgeon-in-Chief, Dr. Stromeyer, whom he constantly accompanied during the campaigns.

Well understanding the imperfections of his work, the Author hopes he may be pardoned, in consideration of the known difficulty of taking proper notes of cases during a campaign, in the tumult of business, and, again, under the depressing influences of the last few months in which it was written.

FR. ESMARCH.

*Kiel, May,* 1851.

# RESECTION IN GUN-SHOT INJURIES.

Injuries of bones form not only one of the most frequent, but also one of the most dangerous complications of gun-shot wounds; their diagnosis and suitable treatment is of the greatest importance, as in numerous injuries of this kind the life of the patient, or at least the preservation of the limb is endangered.

For the diagnosis of gun-shot wounds affecting the long bones, the injuries of the shafts (Diaphyses) must be considered separately from those of the joint-ends (Epiphyses), as from the different structure of these parts, they are not affected in the same way by projectiles—and again with the latter the joints are implicated.

If heavy projectiles strike a limb, the shattering of the bones and crushing of the soft parts is in general so great, that immediate amputation appears to be the only means to be employed.

The circumstances are different, however, if the injuries are due to bullets or similar small shot.

## I. On the injuries of Shafts of Bones by Bullets.

### 1. On the different kinds of these injuries.

It occurs in rare cases that a bullet strikes the shaft of a long bone and pierces it without destroying the continuity. This happens especially in those places where the spongy substance composes the far greater portion of the volume. In the upper third of the tibia three such cases occurred; two of the patients were cured comparatively quick, the third proved fatal after protracted suffering, as the greater part of the tibia became necrosed. In 1848 I treated a patient in whom the left ulna was pierced close beneath the coronoid process, and yet not broken. The Danes employed at that time cartridges containing two small bullets and a piece of lead, with one of which bullets the injury was probably caused, as the two little fingers could only with difficulty be introduced so far, that their tips met in the centre. This case also healed perfectly and without anchylosis of the elbow joint. Hennen relates two cases, in which the femur was pierced in the middle of its shaft by a bullet without fracture: he could introduce a finger through the aperture, which had sharp, clean-cut edges. [In St. George's Hospital Museum is a humerus in which a bullet has lodged in the wide portion of the shaft, just above its articular surface; a little more force would have completed its passage without fracture.]

A more frequent occurrence is, that large portions of the shaft are struck off without causing fracture. Most of these cases are successful, unless from external agents or through unsuitable treatment, inflammation of the bone sets in, which in such case often results fatally in pyæmia. Such a result is especially to be feared, if the bone has been merely bruised by the bullet, without experiencing loss of substance. We have observed many cases where a spent bullet had struck and flattened itself against the femur, and had lodged close behind the same in the soft parts. On examination a portion of the bone was found deprived of periosteum. The bullet was generally discovered and removed after some days. In four of these cases enormous formation of matter took place, and death followed by absorption of pus. On the autopsy, the spot bared of periosteum was found discoloured and necrosed, and the whole medullary canal filled with fœtid pus as far as the epiphyses. Such cases are not rare, and must be met with the greatest attention; the strictest antiphlogistic treatment is necessary so soon as active inflammatory symptoms arise. By this means it is generally possible to check the farther extension of the inflammation, and to prevent the excessive suppuration.

Grazing shots, which tear the scalp and lay bare the skull with destruction of the periosteum, occasionally end fatally in a similar manner; the danger here also depends on inflammation of the bone, which may spread to the veins of the diploe, and then excite pyæmia.

In by far the greater number of cases, bullets striking the shaft of a bone cause actual fractures, accompanied according to circumstances, with more or less comminution. If already spent, the bullet often causes merely a simple breach of continuity, and lodges between the fragments or in the neighbourhood; occasionally in such cases the clothes are uninjured, are driven before the bullet into the wound and then by the movement of the patient the whole may be drawn out without being perceived. The clothes should be carefully examined therefore in such wounds. In the attack on Friedrichstadt, Captain B. had the left femur broken by case-shot in the upper third,—in the lappet of the frock-coat there was a four-cornered hole of the size of a quarter inch, the surgeon had already many times sought for the bullet in vain, causing much agony to the patient. Dr Stromeyer had the trousers examined, and found two parallel rents an inch apart, and neither large enough to allow a case-shot to pass. The stronger longitudinal threads had not given way, and the bullet must have fallen out again immediately. Further examination was not made, and the fracture and wound healed, though with considerable deformity and after protracted suppuration.

A bullet traversing a limb, causes merely a simple fracture, if it only grazes the bone, and does not strike it at a right angle. Its direction is now often changed, and a line drawn straight from the entrance to the exit of the bullet would seem not to have interfered with the bone. On examination with the finger, however, from the entrance opening, the broken ends of the bones are usually quickly found.

If cases of this kind are properly treated, and the sick man is not actually in the way of pyæmic influences, the limb may usually be saved with frequently very little shortening. We have seen not a small number of fractures of the thigh of this kind result favorably. The prognosis is bad only when the femur is broken in its upper third, or in the neighbourhood of the trochanters, as in these cases the thickness of the soft parts, and the impossibility of fixing the fragments sufficiently, tend much to the occurrence of severe inflammation and fœtid suppuration.

The kinds of injury to bone hitherto described are of the rarer occurrence, hence one must be careful, lest after slight examination a favorable prognosis is given on the supposition of such a state being present, while overlooking the more common extensive injuries.

The bone is usually broken in many fragments where struck, and besides this longitudinal fissures frequently extend far upwards and downwards in both fragments of the shaft.

## 2. On Comminution and Fissuring of bones.

The number of splinters is very different, three or four may be found, often however thirty, forty or more. Their size is necessarily equally various. The smaller number of these fragments preserve their connection with the soft parts, it is therefore wrong to consider that the whole act as foreign bodies and must be thrown off. Many observations have shown us that such splinters have not lost their vitality, if they by means of the periosteum are still connected with the soft parts, but that under favorable circumstances they will become again united with the fragments of the shaft and with each other by means of the callus.

A preparation of Stromeyer's proves this assertion most clearly. It is the humerus of a Schleswig-Holstein soldier, in whom at Idstedt both the left femur and humerus were shattered by musket-bullets. At first the case proceeded favorably and the humerus was consolidated in the third week without removal of a single sequestrum; the femur also seemed inclined to heal although the bullet remained in the wound, however, in the fourth week the discharge became sanious, pyæmia developed itself from this time and proved rapidly fatal. During the last few days the fracture of the humerus again gave way; on the autopsy it was found that this bone in its middle was in eight large splinters, of which five already were firmly united to the lower—, three to the upper—fragment, by means of a considerable mass of callus. This was evidently re-absorbed at the spot where the splinters had been mutually united, and thus the continuity had been again broken. The femur, simply-broken close beneath the trochanters, showed similarly at its fractured ends a copious deposit of callus, and many fissures which extended far into the bone were so filled with the same, that they could hardly be recognised as such. The fractured ends were necrosed on account of the extension to them of the foul state of the rest of the wound.

Dupuytren names those splinters which in the first instance still remain in connection with the soft parts—secondary splinters, and says of them, that they are at a later period thrown off and discharged by suppuration. This dogma is incorrect and must be so much the more seriously opposed, as it has led military surgeons to operative interference, in which in our opinion they are not justified. For instance Baudens, in his "Clinique des plaies d'Armes a feu," has given as a rule that in comminution of the bone by gun-shot, not only all the splinters loose or not, must be removed, but also the broken ends of the bone are to be sawn off above the comminution.

It is an intention of this essay to show that such a treatment is in the majority of cases is unnecessary, in many directly hurtful; if in the commencement of the war this rule was at first followed, yet experience has at last entirely denounced it.

It will be necessary to put aside Dupuytren's arrangement of sequestra, on the one hand as being merely accidental and otherwise as leading only to malpraxis.

Dupuytren describes: 1. primary sequestra, which are at once separated by the shot from the bone and soft parts; 2. secondary sequestra, which still remain connected to the bone and soft parts by tendinous, muscular, ligamentous and similar bonds, and by means of suppuration are thrown off at variable periods, in eight, ten, fourteen, twenty days, a month or even later; 3. tertiary sequestra, which arise in consequence of the contusion of bone caused by the projectile in the neighbourhood of the fracture, and which are due to a peculiar natural process, usually protracted perhaps for ten, fifteen or twenty years.

It is certain, that the primary and secondary splinters are not essentially different and that they can be only with difficulty distinguished in practice. It depends on circumstances, whether those entirely free shall be removed at once after the injury or in the further progress of the case, and whether those still connected to the soft parts, shall become loosened by suppuration, or unite with the remaining fragments of bone. It would be better merely to distinguish the two kinds, 'fracture-splinter' and 'necrosed-splinter':—they are easily known by their form. I name those fracture-splinters which are by the force in action entirely separated from the bone, whether they still remain united to the periosteum, or not. They are distinguished by their sharp edges from the necrosed-splinters, which have jagged, uneven borders. These arise in consequence of inflammation which has its seat in the natural chasms and canals of the bone, and here effuses its products; by this effusion the nutrient vessels are pressed upon, and thus greater or less portions of the bone caused to die; after being raised from the bone beneath by the formation of granulations, these loosened portions show that jagged, gnawed contour, due to these chasms and canals, while the contour of the fracture-splinters is generally as sharp as broken glass. Of course fracture-splinters, at first retaining their vitality, may become necrotic and they are now difficult to be distinguished from others. Generally such a distinction is unnecessary—the nature and progress of the process should be understood, however, as thus alone a rational treatment can be adopted.

It is decisively agreed that the bone chiefly draws its nourishment through the periosteum.

On considering in what manner the comminution of the bone occurs, it is very evident, how the majority of splinters—especially the larger ones—remain in connection with the periosteum. The bullet acts as a kind of wedge which is driven through the bone; that part of it which it directly strikes, it drives before itself and crushes into minute pieces, which usually hang about the walls of the shot-canal near its exit from the body. While the bullet forces its way through the bone, it compels this brittle body to yield in all directions, and thus arise a number of clefts which isolate many fragments entirely from the bone. One easily sees, however, that the force is centrifugal and that the periosteum is exposed to no further tearing when once the bone is splintered. It merely tears in those places corresponding to the clefts, but is not separated from the outer surface of the splinters. The whole of the soft parts surrounding the bone suffer a not inconsiderable crushing, as they are enclosed between the skin [and fascia] and the spreading portions of bone, and to this crushing of the soft parts I think we must refer many of the symptoms immediately consequent on gun-shot wounds. If a lower limb is struck at the moment when the weight of the body bears upon it, the injury will be much more extensive, for the upper fragment is thrust down by the whole weight of the body into the soft parts, the splinters separated, and from many of them the periosteum torn off.

Those splinters which have lost all connection with the periosteum, lose their vitality, and, acting as foreign bodies, must be removed sooner or later.

Occasionally the splinters are yet firmly united to the muscles by the tendons, and much time must elapse before they become loose. Not unfrequently those entirely separated at the time have been simultaneously driven into the medullary canal, or remaining between the broken ends, they hinder the complete closure of the wound; in favourable cases, such splinters may become encysted, as bullets or other foreign bodies may be,—usually, however, they excite fresh inflammation and formation of matter, and if not removed at the right time, cause so much annoyance to the patient, that he frequently wishes that the limb had been rather amputated. An operation for necrosis is frequently now beneficial.

I conceive that the tertiary splinters of which Dupuytren speaks are to be referred in many cases to these primary 'fracture-splinters' incarcerated by callus. Although I do not deny that the contused ends of fragments in gun-shot wounds often become necrosed, yet experience has taught us that it is not so common of occurrence under favorable circumstances and with suitable treatment as has been expected, and the removal by the saw of the fractured ends is to be entirely given up, unless necessitated by peculiar circumstances; as, when nerves are irritated by sharp points of bone.

The necrosis occurring after comminution of the shaft, is not a consequence of the concussion, but of inflammation of the bone with suppuration, healthy or unhealthy, this may equally well arise upon a flat sawn surface as upon the surface of the fracture immediately caused by the bullet.

Similar principles hold good with respect to the fissures occurring very often in comminution of long bones, and extending themselves frequently upwards and downwards for a great distance. The older surgeons have considered these fissures as very dangerous, and they are so, if the suppuration becomes foul and the bone inflames. In such cases the suppurative action follows the course of the fissures and extensive necrosis is the result, unless the patient is carried off by the phlebitis in the bone common in such cases. Under favorable circumstances, however, these fissures heal just as well as fractures, by means of effused callus, which shortly removes every trace of injury. Fissures of bone may equally well arise in simple contusions or injuries, in which no actual fracture is present.

If the fissures extend into the joint, the injury is of course much more dangerous. The study of the preparations, however, collected by Stromeyer, in the campaign, evidences a most important practical point, that in comminution of the shaft of a long bone, the fissures almost never extend into the epiphyses; in the same manner injuries of the epiphysis only, in extremely rare cases extend into the shaft, unless the bullet strikes the adjoining borders of both parts, in which case both are usually more or less seriously comminuted.

This is certainly due to the soldiers being generally young, and the parts not yet consolidated. This is seen on making a section—actual cartilaginous substance may commonly be found, and on maceration the parts separate, as we found, in many instances. The abrupt termination of the injury is therefore well explained.

It is well known that a similar state occurs in the skull, the fissures not extending beyond the sutures in young individuals, if the force is of limited effect, while in older people the fissures may extend through many bones.

### 3. On the Cure of comminuted Bones without operative Interference.

I have shown that extensive comminution of the larger osseous shafts may be cured without operative interference; it is evident that for this, favorable external influences and proper treatment is necessary. It was long known that extensive comminution of the facial bones and upper jaw heal not unfrequently without the discharge of any considerable bony fragments. We therefore left all these injuries almost entirely to nature, and had no cause to be dissatisfied with the result. In the injuries of the bones of the fore-arm we made similar trials, and thus emboldened, even in extensive comminution of the humerus we ventured to trust to nature. The greater number of cases were successful, hence we adopted the rule, not to perform primary amputation in cases of comminution of the humerus, unless it was demanded by other complications.

We are able to show a certain number of cases where the femur was comminuted, and in which recovery took place without any operation.

In the most favorable cases of this kind the wounds heal without great suppuration or discharge of splinters. Usually it is necessary to extract the sequestra in the course of the case, whether by the track of the projectile, or by incisions, rendered necessary for the discharge of matter. The fractures consolidate themselves then in from six to ten weeks, and after all foreign bodies—splinters, bullets, pieces of clothing &c. —are removed, the cutaneous wounds also close— the wound of exit usually first; by means of suitable exercise the use of the limb is recovered.

### 4. On Circumstances exerting hurtful Influences on the Wound.

#### a. Influence of Transport from the Battle-field.

The wounded man is exposed to many noxious influences, which no care of the surgeon can ward off. The first, and that the most hurtful for injuries of bone, is the transport from the field to the hospital, generally by rough roads and imperfect means of conveyance. An army can never take with it proper vehicles in sufficient number for the wounded; common waggons with straw must then be used and it is fortunate if there are enough. By suitable bandages some of the evil consequences to shattered limbs from the jolting of the vehicle may frequently be avoided, but time and quiet are necessary, and neither can be obtained. Every one of our surgeons knows how difficult it was more especially in our hedge and ditch country, to give proper attendance to individuals, at the same time that certain bodies of troops were compelled to retreat.

The phenomena accompanying comminution of a large bone are those of severe concussion of the limb. I have already explained how the soft parts are crushed by the violent flying apart of the pieces of bone. In consequence of this, more or less hæmorrhage occurs in the cellular tissue of the surrounding soft parts. The sheaths of the vessels and nerves are especially the seat of hæmorrhagic effusion. Lessened sensibility [benumbing] is the next result from contusion and compression of the nervous trunks, and the same cause acting on the veins, the circulation is slackened and passive congestion arises, shortly followed by copious serous infiltration in the tissues.

This Serous Infiltration forms a white, doughy, painless swelling, which at first only occurs opposite the injured part of the bone. A bandage quickly and well applied and by which the whole limb is enveloped, and the application of splints or straw pads (straw splints) to check the motion of the fragments of bone as far as practicable, may contribute much to check or prevent this serous infiltration. If such a bandage cannot be applied, the transport of the wounded exerts the most hurtful agency on these injuries. Every jolt of the waggon causes severe pain, the muscles are thrown into spasmodic action, so as to drive the sharp points of bone again and again into the soft parts. It is especially hurtful if a careless surgeon has applied a single or a few turns of a bandage or indeed a tourniquet firmly around the injured spot for fear of hæmorrhage. The congestion and effusion increases more and more, so that on arrival at the hospital, the bandage may be buried in the swollen limb.

Such a mistake is still worse should large vessels, especially veins, be wounded, the hæmorrhage is aided, not checked, and distends the whole limb, as the outer wound is closed; it forms a firm swelling, the skin over which is at first pale and cool but soon becomes of a spotted, dirty, brownish color in the course of the veins, later blebs appear on these spots, containing dark red serum.

This Bloody Infiltration generally results in mortification, if it is impossible to remove a portion of the stagnant fluids by suitable treatment, and thus to restore the circulation. On dissection of such limbs, thick layers of black, coagulated blood are found in the cellular tissue, under the skin, between the muscles and within the sheaths of the vessels and nerves. Of course where no blood is found the distension has been due to serous effusion.

### b. Influence of Hospital Air on the Wound.

It is well known that in hospitals where suppurating wounds have been long under treatment, and where ventilation has not been properly attended to—that miasma developes itself and exerts a noxious agency on the wounds. Those wounds complicated with injuries to bone are especially subject to this influence, and in such cases phlebitis resulting in pyæmia with generally a fatal termination may readily occur in wounds of apparently a slight character. If the hospital air is very foul, hospital gangrene is set up, it is contagious, affecting the merest wound and often proving rapidly fatal. In the Schleswig-Holstein campaigns we have happily never experienced this evil,—pyæmia, however, was very fatal, especially in those sick-houses which had been repeatedly filled by patients with wounds of a severe description. Thus, for example, in 1849, the roomy Moravian church at Christiansfeld was employed as a hospital, and on April 23rd, was filled with soldiers severely wounded at Kolding, the treatment was successful, many were healed or were sent to the south convalescent. Several severe cases remained there the whole summer, and although few in number, prevented a thorough cleansing being effected. After the attack on Fridericia, on the sixth of July, this hospital was again filled with cases of severe injuries, and the sad results soon showed themselves. For pyæmia took off so many and so quickly, that the surgeons were in despair; for instance, all but one case of amputation of the thigh proved fatal, and even a case of amputation of the arm, transferred from Kolding, in whom the cicatrix was nearly closed and who could already leave the bed and chamber, was attacked and sank under pyæmia.

This is not the place to consider the nature of pyæmia, we generally found its cause to be inflammation of the veins. It is especially the veins of the bones from which this affection proceeds, for when we made longitudinal sections of them in cases of injury to bone fatal by pyæmia, we generally found the medullary canal filled with sanious pus. In many cases we found the veins issuing from the foramen nutritium in like manner inflamed and filled with pus, the suppuration, however, seldom extended higher in the large venous trunks.

### 5. On the Inflammation and Suppuration in the Wound.

Soon after the reception of the wound, inflammatory symptoms arise in the injured parts; under favorable circumstances this leads to healthy suppuration and granulation and more or less rapid healing of the wound, after the loose splinters and other foreign bodies have been removed. This inflammation is necessary for the discharge of the crushed cellular tissue with the pus, the duty of the surgeon is, to retain it within proper limits. In unfavorable circumstances it brings the patient into the greatest danger; if, for instance, the wounded part was much infiltrated with serum, the inflammation increases the stagnation, fibrinous exudation follows, the swelling becomes firmer, rapidly proceeds beyond the nearest joint (as in injury of the humerus, over the shoulder), the skin becomes glistening, red, excessively distended, and shows occasional blebs. Directly the suppuration sets in, it spreads most rapidly through all the tissues previously infiltrated, numerous deposits of matter form in all directions, giving enough to do at once to the patient and the surgeon, (Purulent Infiltration).

In certain cases the inflammation progresses by the congested smaller veins and by the lymphatics of the limb, the distended skin then shows a deep erysipelatous redness, the patient has violent fever, a dry tongue and becomes delirious. Difficulty of respiration follows and death in a few days. On the autopsy, the dissection of the limb gives oozing of pus in numerous points, clearly from the mouths of the smaller vessels. Usually lobular pyæmic abscesses are simultaneously found in the lungs in the first stage.

A previous hæmorrhagic infiltration extraordinarily increases the severity of the inflammatory symptoms; if large vessels are wounded and the extravasation is extensive, merely a slight inflammatory swelling is necessary to check completely the circulation in the limb; the member below the wound becomes blue, ice-cold, insensible and gangrene is not to be prevented. In a few bad cases, indeed, no gangrene occurs, but as soon as suppuration sets in, the coagulated blood putrefies, the suppuration becomes ichorous, and as the injured bones are bathed in this fluid, the danger of phlebitis in the bone becomes so much the more imminent.

For the formation of callus, a certain hyperæmia and exudation of plastic material is necessary. Every thing depends on holding the inflammation within bounds until the callus is thrown out, if the inflammation is too severe, the plastic material changes into pus, this fills the passages of the bone, compressing the vessels and thus in favorable cases causes necrosis of the fractured ends and of the splinters still adhering to the periosteum. These sequestra keep up suppuration and may threaten the life of the patient by mere exhaustion. In worse cases, however, the veins of the bone are likewise inflamed, phlebitis occurs, (as previously described) and the result is generally fatal through pyæmia. Even when callus is formed and consolidation effected, a bad sanious discharge may destroy everything—the callus is absorbed and the fragments become necrosed.

Such gun-shot wounds in which, besides comminution of the bone, other complications are also present; as, injury of the great vessels, opening of the joints, &c., become in this manner so much the more dangerous, that it is often necessary, if hæmorrhage or inflammation of the joint sets in, to remove the limb, although the injury to bone would of itself not have required such interference.

Trismus and Tetanus, we have seen but rarely after gun-shot wounds, in which, however, most commonly the bone was also injured. The acute cases all proved fatal in spite of treatment, the chronic cases were successful under the use of warm baths and moderate doses of morphia.

## 6. On the Examination and Diagnosis of the Injuries.

The practice of the surgeon cannot be rational, unless he has followed the disease under favorable and unfavorable circumstances. For the right diagnosis and treatment of special cases, he must likewise possess an acute practical observation, power of diagnosis exercised by experience, and a decided method of treatment, constantly pursuing the object in view. These characteristics are especially requisite as regards examination of the wounded, who are often brought before the surgeon unexpectedly and in great numbers. On an exact diagnosis the life of the patient frequently depends. The surgeon must therefore determine the kind of injury, as soon as the patient is brought into the hospital. Many wounds require no examination, as the mere external aspect informs the skilful surgeon of the whole state; in other cases it is of the greatest importance to perform an examination, especially if it is doubtful whether an operation is necessary, or not. Immediately after the injury the wound may be examined without much pain, as the finger can enter where the bullet has passed. As soon, however, as the parts are somewhat swollen, the examination becomes very painful for the patient, causes bleeding and allows the air to enter to the wound. Nevertheless, in important cases an accurate examination must not be omitted, and in no case does the surgeon trust to the statements of the patient, or to another surgeon, whom he cannot implicitly confide in.

If the question is, whether the limb should be preserved, or not, the entrance of the bullet-track must be dilated by the knife if required by the extent of the swelling; examination by the probe is of no use, as it gives very scanty information. Ambroise Pare's advice, to place the limb in the position in which it was when wounded is of the greatest importance and frequently obviates the use of the knife; unfortunately, however, the patient cannot always say what position the limb had occupied, hence during the examination it must be put in various directions.

It is not possible to cause crepitation at all times, even when the bone has been extensively comminuted, and occasionally the patients themselves can perform all the motions of the injured limb.

If it is decided to preserve the limb, those splinters entirely loose may be at once removed by the finger, care must be taken not to search too much, as more evil may be caused than there is relief afforded by the withdrawal of the splinters.



### 7. On the Treatment of Wounds.

If on the examination an injury is found which, either on its own account or from the circumstances of the patient, renders preservation of the limb impossible, amputation should be proceeded with forthwith. The great benefit of immediate amputation is long proved by writers on military surgery. It was found by experience in Schleswig-Holstein, that the danger to life increased considerably with each hour of delay; hence amputation was performed as early as possible, if sufficient indication existed to call for it.

I speak now only of the practice in the hospitals. Very few amputations were performed by the surgeons who accompanied the troops to the field. This is accounted for by there having been only one chief ambulance in the first two campaigns, and that though well supplied, was unwieldy; hence it could never be brought near a battle, lest it should hinder the troops, or be left in a retreat in the enemies' hands. The surgeons were therefore obliged to depend pretty much on themselves.

Dr Stromeyer in 1850 introduced the Brigade-ambulances(*) which corrected this evil, by a number of waggons being collected in suitable spots. Yet as of late the field was not far from hospitals, the severely wounded were generally sent off at once,—this circumstance accounts for the successful operations at Schleswig after the battle of Idstedt. It was only at the attack on Fridericia that many severe operations were performed at the brigade-ambulance at Suederstapel, about three miles distant: the result of these operations was not especially favorable, as the transport to the hospital was over very bad roads.

If there is hope of saving the limb, the treatment must be such as corresponds to the previously described course of the case. Before all therefore care must be taken that no further injury is experienced by the patient. No motion should be allowed of the fragments of bone on one another, and thus pain is prevented being caused which should excite the muscles to renewed spasms.

* Compare. Armee und Militairesanitaetswesen der Herzogthuemer S.-H., von Dr. Adolf Erismann. Bern. 1851.

Large cushions filled with chaff, or chopped straw are of the most use. If the limbs are much shortened by over-riding of the fragments, no extension is to be employed as yet, for the muscles will be merely irritated with no good effect; after some days they yield spontaneously under suitable treatment, and it is easy now to bring the limb into a better position by the use of splints.

If the humerus was comminuted, a cushion was laid between the arm and chest, and then by means of a sling and roller the arm was retained close to the body;—the fore-arm was merely laid upon a flat well-padded splint, having a foramen for the internal condyle of the humerus. Fractured thighs were laid either on a double-inclined plane or simply upon good-sized cushions in an abducted and outwardly reclined position; in injuries of the leg, Heister's fracture-box was very useful.

The arrangements are better as they are more simple, generally large cushions alone are employed, half-full so that the contents may be moved and the cushions take various forms. Patriotic ladies often furnished little cushions which the patients enjoyed, but we found a large one answer the purpose better.

The greatest care must be given to put the limb in good position, and as a rule the bed of the patient must not be left till he is free from pain, it occupies more time but rewards one better in the favorable consequences, not to speak of the thankfulness of the patient. A firm splint apparatus must not be used until all irritation has ceased, and the suppuration is declining. It is then always time enough to alter the mal-position gradually and to bring the bones into a better position.

Where there are two long bones in a limb, so much care is not required if one alone is injured,—the other one acts as a kind of splint.

It is a not less important duty to take every care to promote free ventilation and thus obviate the production of miasma. In general only such buildings should be used as hospitals, which are lofty and free, but often any must be employed for the accommodation of the wounded. The surgeon can still do much here, but he must himself see that his directions are followed. Ventilation can always be enforced, draughts must not be feared, and a window always open. Patients and nurses fear the cold, and shut the windows if they no longer expect the visit of the surgeon; he must however look in at various unexpected times and enforce this point,—it is perhaps best to remove a portion of the window. We were at first in great fear of draughts on account of the prevalence of rheumatic disease, however on becoming bolder, no evil consequences ensued.

Care with regard to cleanliness is equally important, especially when, as with us the worst soldiers were employed as nurses. The surgeon must look after all, and punish the least want of cleanliness. In the first two days after a combat this cannot be carried out, as the surgeon has otherwise far too much on his hands. If as Heister relates, many patients died after the battle of Leipzig from want of cleanliness, we must not charge this upon the medical men, as they were far too few in number for the duties they had to perform.

In dressing the wounds, cleanliness is of the last importance. The foul dressings should be at once sent out of the room, not carried from bed to bed, and either a separate sponge employed for each patient, or the wound should be cleansed with dry charpie or lint after trickling warm water over it.

In the treatment of the wounds we strove for the greatest simplicity. Certain points here are still under discussion ; such as the dilatation of wounds by the knife, the extraction of splinters and foreign bodies and the employment of general depletion. Some of these questions have been already considered, and now without further delay, I shall describe the treatment adopted by us as the best in the last campaign. The surgeon has only to keep within bounds the the pathological processes of the wound , and to remove impediments to its cure. The first demands antiphlogistic treatment, the second—the removal of foreign bodies. I have already shown that the removal of foreign bodies can be more easily effected at a later period, and that by no means all the splinters of bone are to be considered as such.

When therefore the wounded were brought into the hospital, and the larger long bones had been comminuted in the shaft, if the limb had been well bandaged and splints applied in the field, we did as little as possible, employing easy position and cold applications, and making no attempt to extract foreign bodies,—it must be understood, however, that the external aspect had already shown that neither considerable infiltration nor any other dangerous complication of the wound was present.

With such treatment we have seen even fractures of the thigh by musket-ball, heal in a proportionably shorter time and without considerable suppuration.

In cases requiring a close examination with the finger, we withdrew at once all foreign bodies and those splinters of bone entirely loose which could be removed without difficulty by the track, but taking care not to cause fresh irritation by a protracted search. Such a search would cause more damage than the gain from the removal of the foreign bodies, as they are in general easier to remove after the occurrence of suppuration. If the examination was requisite and the wounded part much swollen, dilatation was employed to facilitate the extraction of the foreign bodies. Dilatation, however, merely because the wound is by gun-shot, or because strangulation is feared later, is not worthy of the surgeon, who should never operate without a decided motive, or if he can reach his aim by other less painful means.

If much infiltration was already present we did not delay in using general depletion before the occurrence of inflammatory symptoms, in cases where cold did not decidedly lessen the swelling, Some surgeons employed instead, long incisions through the integument and fascia and arrived at the same end certainly quicker by the loss of the serous infiltration and of some blood from the incisions.

Venesection is, however, equally efficacious and thus the pain of incisions and their subsequent suppuration may be spared.

Only where serious hæmorrhagic infiltration is present, one is sometimes necessitated to employ such incisions, as the tension is thus quicker overcome, and in such cases every delay increases the risk of gangrene.

In very many cases, however, the serous infiltration subsides by the mere application of cold, that is to say, when ice is procurable.

In the year 1850. after the battle of Idstedt, we had in Schleswig plenty of ice at our disposal, so as to be able to employ it in all cases of comminution of bone and indeed we found the most excellent results from it.

If no ice could be procured, cold water dressing was employed, acting also as an antiphlogistic but less powerfully. Under these circumstances the wound must not be closed with charpie or adhesive plaster, but the wet cloth immediately applied to it, and not covered with impervious cloth as then it would become warm, and when reapplied, the continual change from heat to cold would irritate and increase the congestion.

In wounds of the posterior parts of the body in order to prevent the applications becoming warm by being laid upon, we have had patients who have lain prone six months and more.

It is to be understood that low diet and frequent saline purgatives must aid the local antiphlogistic means.

The use of cold is not necessarily given up on the occurrence of suppuration, on the contrary in wounds of joints the ice-bag was used for six to eight weeks and longer, and in certain cases with the best success.

In general poultices or warm applications were employed from the occurrence of suppuration on the fourth or fifth day, usually the change was made at the desire of the patient; the greatest simplicity was sought for by us in this matter also, being in fact forced to it by circumstances. In the first campaign poultices of oatmeal, &c., were employed and occupied much time in their application. In 1850 in Gottorp, we were obliged to satisfy ourselves with warm water dressings and found them answer admirably, and perhaps better than the poultices. They are cleaner and one nurse may fill a pot for each patient every two or three hours, so that he can dress his wounds,—the wet linen is covered by oil-cloth, oil-silk, or gutta percha. Charpie must not be used to cover the wound, as a chief use of this warm-water dressing is to allow the pus to flow from the wound at each renewal of the dressing.

If purulent deposits were threatening to develop themselves in the hardened parts of the serous infiltration, local depletion would frequently check any further progress. Some surgeons made early incisions, but we often experienced the same service from a few leeches. In many cases the inflammatory swelling proceeded to suppuration, especially if sequestra were present in the wound.

Pressure to evacuate the matter by the wound, instead of employing the knife, repeated each time of dressing by some practitioners, is often of no use and only causes irritation. The collections of pus often become more copious and its quality degenerates, for the inflammation is increased each time, and the violence causes oozing of blood, which corrupts and pollutes the pus.

In such cases we made free incisions, or dilated the wound so as to reach the collection of matter. If such incisions are undertaken they must be large enough to introduce the finger for the extraction of sequestra, &c.,—in most cases this can be done. Prudence forbids protracted searching or violence. Of course the incisions are such as allow the free discharge of matter; by their suitable position and direction and by their timely employment, one may recognise a good surgeon.

In the last campaign we rarely employed poultices, unless for injuries of the hand or of the foot, and then with great service. Hot baths for the hand and foot are of equal importance, but are not to be used until the inflammatory swelling is nearly subsided. They promote the discharge of matter remarkably, and seem useful by exciting activity of the cutaneous vessels. Dr. Stromeyer had tin pans made for this purpose, provided with loose covers from which a portion was cut out to allow the passage of the limb. When filled with warm water and covered with a woollen cloth, the heat was long retained. They were nine inches high and broad, and two feet long.

As soon as the suppuration diminishes by discharge of the sequestra and cleansing of the wound, flannel bandages are used, and rapidly remove any œdematous swelling. On account of the moving of the limb requisite for this purpose, the fracture should be first well consolidated, if not so—the many-tailed bandage proves very useful, its pieces being passed under the limb by a spatula without raising it from the bed.

The dressing for wounds become superficial and in which cicatrization was ready to begin, was a solution of Nitrate of Silver 1 — 5 grains in the ounce, on linen rag covered with oil-silk. Stromeyer used no ointments, except occasionally charpie dipped in oil, especially in those cases where the patient could rise, as warm applications might have exposed the parts to be chilled.

These are the chief principles of treatment which seemed to us the most desirable from the conflicting experience of three campaigns. The practice differs much from that followed in the commencement of the war, it would not however, have been creditable if so many observers had not advanced our knowledge a good deal; yet many questions remain undecided.

Up to 1848, Baudens was the most modern author on Gun-shot Wounds. It was natural therefore to look upon him as an authority, so that following his rules, in that year, besides many resections of joints, resections in the continuity of bones were frequently performed by the Surgeon in Chief Dr B. Langenbeck or under his direction. The results were to be considered favorable rather than not so, and without doubt many lives were saved, and limbs preserved which would otherwise have been amputated.

In the second campaign, however, Dr Stromeyer, who had in 1848 treated many gun-shot wounds at Freiburg, adopted the opposite plan and did not perform primary resection in the continuity. Far fewer operations were therefore practised,— in many cases we contented ourselves with extracting splinters after dilatation, in others we did nothing of the kind and here the results were so favorable, that in the last campaign no single resection in the continuity was performed.

It was objected to Dr Langenbeck that he operated for resection in the continuity too frequently, the then stage of military surgery, however, and the assertions of Baudens, required the trial to be made.

On Langenbeck succeeding to Dieffenbach's chair, he lost the opportunity of further experience which doubtless would have conducted him to the same result.

The majority of resections performed by B. Langenbeck were of other kind than primary. It could not be otherwise, as from the want of experience of the surgeons, or from the noxious surrounding circumstances which at that time could not be removed:— he frequently found patients, in his visits to the hospitals, with their wounds in a neglected and bad condition. By the foul suppuration the yet-attached sequestra and the ends of the fractured bones were become necrosed, and the patient so reduced by the severity of the fever and excessive suppuration, that the surgeon in charge had frequently determined on amputation. Langenbeck preserved the limb in many such cases, by dilating the wound, removing all loose sequestra and sawing off the necrosed ends. This must not be confounded with primary resection — the periosteum could easily be preserved by separating it, in its now thickened condition, from the piece to be removed. In other cases only such portions were removed, whose preservation was hopeless and which had a hurtful influence on the wound.

Unfortunately, I was not able to see the favorable cases of this kind, which occurred at Schleswig and Rensburg after the battle of Schleswig in 1848, the cases later from Hoptruep and Dueppel supplied many similar operations at Flensburg, of which a part were unfavorable. The air and locality, however, were unhealthy, so that many wounds of much less severity proved fatal at the last-named place, and the result is therefore not decisive. I often observed the great improvement in the character of the wound and its secretion and in the state of the patient consequent on the operation, and that the loss of bone was, after entire cicatrization, more or less quickly restored by the formation of callus.

Our increased experience gave us more and more insight into the treatment of wounds. It must be chiefly ascribed to the better placing of the limb, more careful antiphlogistic treatment and attention to the treatment of purulent deposits,— that the degeneration of the healing process in wounds was much rarer in the last campaign. When occurring at this time, being generally a symptom of pneumonia, it forbad operative interference, as we found this merely hastened the fatal result. If arterial hæmorrhage occurs, amputation must be performed instantly, as ligature of the vessel in cases of severe injury of the bone, has either no result or induces gangrene. The so-called parenchymatous hæmorrhage, proceeding from the capillaries and smaller veins, is due to obstruction in the larger veins, and can neither be stopped by ligature of the artery nor usually by amputation, as the pyæmic process is already in progress.

Statistical results from the surgeons in Schleswig-Holstein would have shown positively, that the indications for operations were entirely different at the end compared with the beginning of the war. Unfortunately no list is at hand of the first campaign, the results of the two last are sufficient to authorize the the treatment I have described.

In very considerable comminution of the shaft of the humerus, amputation was not rarely performed in the two first campaigns. In 9 cases, preservation was attempted by removal of the splinters and resection of the ends of the fragments; 4 of these patients died, and of the remaining 5 many retained a very defective limb. In 7 similar cases in 1849, consolidation was essayed without resection, by immediate removal of the loosened splinters in 3 cases, and after the occurrence of suppuration in the 4 others. The result was beyond expectation, as but 1 of the 3 first proved fatal. and in the 4 last the the recovery was complete and comparatively rapid. In 1850 therefore, in such cases we followed the same (last mentioned) practice, and with surprizing consequences. Of 25 cases but 4 died, in the remainder a complete cure followed, although in many the humerus had been shattered by cartridge shot. In all these cases the fracture was fully consolidated, and in many the usefulness of the arm was almost entirely restored. In one of them, the humerus had been shattered, at Fridericia, in its upper third, upwards of sixty sequestra were discharged, and yet complete consolidation took place with the loss of but two inches in length, and the movements of the arm were already powerful, when the patient left the hospital.

Comminution of the shaft of one or of both bones of the fore-arm does not of itself indicate operative interference, other complications may necessitate primary or secondary amputation. This is proved statistically. The resection was undertaken in 6 cases in the first—in 1 case in the second—campaign. In 1849, all splinters, whether free or attached, were removed in 7 cases of comminution of one or both bones, without resecting the fractured ends. Recovery was much slower and less complete in these 14 cases, than in those left entirely to nature in the last half of the second—and in the third—campaign; of 41 but 1 died and he from Asiatic Cholera, the remaining 40 are entirely recovered, of whom 6 had injury of both bones, 16 of the radius, 18 of the ulna. False joints I have not seen after the expectative treatment, but many certainly after resection or after the primary extraction of all the splinters, and especially after injury of one bone, probably because the sound bone prevented the requisite approximation of the broken ends, so as to effect consolidation.

In comminution of the shaft of the femur, amputation is generally unavoidable, especially if the injury to bone is extensive or if the soft parts are much lacerated and contused. If the fracture is without comminution, or this is not extensive and there is no complication of the soft parts—preservation of the limb must be attempted. One must not try to remove at once all the splinters from the wound, for the irritation and admission of air increases the inflammation. By prudent treatment we were able to heal 12 out of 26 cases of this kind, a result which may be considered very favorable, as from the whole number of amputations of the thigh in the three campaigns, 3 out of 5 were fatal, only 51 recovered in 128 cases. The resections in the continuity of the femur, essayed in only three cases, were all fatal.

In comminution of one or both bones of the leg, the attempt to preserve the member must always be made, unless hæmorrhage or other complications demand its removal. If this is not the case healing may be expected with suitable care. Resection, or too early removal of the splinters can easily become hurtful, as the closely neighbouring arteries are readily wounded. Of 13 resections in 1848-9, (of which 3 were on both bones, 7 on the tibia and 3 on the fibula,) 7 proved fatal. Namely, 2 of those on both bones, 4 on the tibia, and 1 of those performed on the fibula. The result of later treatment was more successful; of 58 of this kind (8 being of both bones, 27 of the tibia and 23 of the fibula,) 52 recovered and but 6 died. Of the fatal cases, 1 was of injury to both bones, 2 to the tibia, 3 to the fibula.

In comminution of the smaller shafts of the hand and foot, operative interference is rarely necessary. It was evident at least in our climate that the excessive fear of tetanus after such injuries is unfounded, as this frightful disease was rare and then more often after injuries of other parts than of the hand and foot. Another point is, whether operative interference will quicken the recovery, from our experience the expectative treatment is far preferable, and that no operation at all should be undertaken unless immediately after the injury.

After such injuries a considerable inflammatory swelling very quickly sets in, and any interference increases the inflammation, as it is necessary to operate in inflamed parts, and the inflammation becomes much more severe and the cure more protracted after such amputation or exarticulation, than when left to nature. We found also that injured parts, at first looking very unpromising, became more or less useful with suitable care; while on the opposite treatment, as in exarticulation of a finger—the wound being severely inflamed—, not unfrequently suppuration followed in the sheaths of the tendons, and thus interfered with the later usefulness of the hand. Again, resections of these parts are almost always unnecessary and often dangerous, as hæmorrhage may follow, only to be controlled by amputation of the hand or foot.

The same holds good for the injuries of several phalanges, or metacarpal, or metatarsal bones, as for single ones of them. Such injuries healed without requiring the removal of many splinters in the period of suppuration, and without leaving the hand or foot entirely useless. One of the most interesting cases was the following; at Missunde, an artillery-man had the outer half of the foot, as high as the upper ends of the metatarsal bones, torn away by a grenade-shot; the wound had a very bad aspect, the upper splintered ends of the three outer metatarsi and the second metatarsus simply fractured, were exposed. The question was between Syme's and Chopart's operation. On my proposition, this case was left to nature, recovery followed without special phenomena in a comparatively short time—the skin from the inner side being drawn over by the cicatrization after the discharge of some sequestra. Both the two first toes with the metatarsal bones were preserved and proved of great service.

The great importance of strict antiphlogistic treatment was very evident, by comparison of the two following cases. A dragoon in 1850, had inadvertently shot himself through the hand, the entire contents lodged and had comminuted the three middle metacarpal bones. At first it was treated with ice, in spite of this, depletion having been omitted at the right period, phlegmonous inflammation of the hand and arm followed in the sheaths of many of the tendons, and caused profuse suppuration. Numerous incisions were necessary, and the patient lay four months in bed. Two months later, a similar case occurred, here a copious venesection was employed as soon as the first inflammatory symptoms arose. The progress of the inflammation was checked, and the wound healed so soon that the patient could be discharged much before the first.

I hope to give further on the principles we adopted in amputations and exarticulations, at another time and in a distinct form.

## II. ON THE GUN-SHOT INJURIES OF JOINTS.

The injuries of the larger joints endanger the life so much, that the surgeon is generally compelled to interfere, and remove or lessen the danger. These injuries are very common in war, and their seat and extent show such variety and difference, that a knowledge of them is of the greatest interest to the surgeon.

### 1. On the various Kinds of Injuries of Joints.

In a few cases the fibrous capsule alone is injured by the bullet, the joint-ends remaining safe. As in one case, where the bullet penetrated integument and capsule on the inner side of the knee, and thereupon escaped from the wound again. A grazing shot may act similarly, lacerating the soft parts without affecting the bone. Sometimes the capsule is only bruised at first, and only opens afterwards on detaching of the slough.

If a bullet strikes an epiphysis, (part of whose surface enters into the joint), —it may either penetrate the bone or lodge in its centre, if already spent, or if only a grazing shot it causes merely a depression, or a groove-like loss of substance. In none of the cases is the synovial membrane necessarily injured, but almost always the epiphysis is fissured to the articular surface, and sooner or later, the inflammation and suppuration progress from the wound and affect the joint.

Usually, however, both bone and synovial capsule are simultaneously injured, the bullet lodges among the fragments, or has traversed them, after having comminuted one or more of the epiphyses. In joints near the surface, as the elbow, the bullet may again fall out of its entrance-opening after causing considerable damage; if the wounded man has not himself observed this, the surgeon may often trouble himself ineffectually and long, this has occurred to us several times.

The evil consequences of injuries to large joints are well known, the whole wide extent of the synovial membrane which is inflamed on the admission of air, the mass of fibrous and tendinous parts which are next affected, the large extent of the free articular extremities of the bones — which are but for a short time protected by their cartilaginous investment, for this is quickly dissolved and the bone laid bare in the wound-secretions,—, lastly the peculiar formation of the joint cavity, which allows free escape of secretion in but few cases ——— all these conditions sufficiently explain the dangerous consequences of injuries to a joint. If extensive destruction is caused by the weapon, the danger is so much the more increased, for the splinters and foreign bodies in the joint greatly increase the irritation, and the wide extent of the fractured surface of the spongy portions of bones offers much room for the extension of the inflammation.

If one has the opportunity to examine such a wounded joint soon after the injury, blood — partly fluid, partly coagulated—is found in the cavity of the joint, and extravasated in the porous substance of the wounded epiphysis for perhaps an inch or more in extent around the injured spot, showing how far the concussion of the bone has spread. We have always found this on examination, for this purpose, sections of the bone must be made by the saw. On suppuration commencing in the wound, the extravasated blood is decomposed, the secretion becomes foul, and the inflammation always proceeds as far in the injured bone as the extent of the hæmorrhagic extravasation.

## 2. On the Course of the Injuries of Joints.

From the preceding remarks it is easily understood, why the symptoms are so extraordinarily severe after the gun-shot comminution of a large joint. Usually serous infiltration rapidly sets in, and to this inflammation soon joins itself. The swelling of the limb spreads upwards as well as downwards with the greatest rapidity, the integument becomes hot, glazed, red and often so tense that blebs form on the surface.

The whole limb below the joint rapidly becomes œdematous, probably from compression on the veins, which may occur with the greatest ease in the neighbourhood of joints, from the tumefaction of the soft parts. High fever usually sets in very quickly—the pulse becomes very rapid, the integument burning hot, the tongue dry, great thirst is experienced and often such severe pain, that no rest is obtained day or night and delirium may be excited. The retention of the wound-secretions and increased flow of synovia, which can seldom escape —distend the capsule of the joint more and more, pressure usually forces synovia from the wound, at first mixed with blood, later troubled and purulent.

Every movement of the limb causes the most frightful pain, perhaps scarcely a touch can be borne. A severe rigor frequently announces the occurrence of suppuration, and sometimes pyæmia rapidly follows, as the great extent of the internal wound-surface is very favorable for the absorption of pus. In such cases the pus soon becomes foul from the decomposition of the extravasated blood, fœtid gas is generated in the joint and mixed with sanies escapes on pressure with a bubbling noise; the patient now falls into raving delirium, or into stupor, the tongue is dry and dark brown, jaundice is now common, and death follows in a few days. Such is usually the course of injuries of the knee, if the limb has not been removed sufficiently early. Amputation only hastens the termination, if inflammation of such severity has already occurred.

If the inflammation is less or affects a smaller joint, the reaction occurs with less apparent severity. In cases where only the capsule has been injured and that slightly, no blood is extravasated into the joint and the pus forms gradually and is of good character; as it cannot, however, escape, it distends the capsule, until it breaks through that portion of it which offers least resistance. Spreading upwards and downwards in the cellular tissue (superficial and deep) of the limb, it excites fresh phlegmonous inflammation—rapidly passing to suppuration—and forming those abscesses often called 'gravitating abscesses', although they may be directly opposed in their extension to the force of gravity. A whole limb may be thus undermined and if pyæmia does not occur, yet the great loss of secretion alone, endangers life and often necessitates amputation.

These symptoms arise very slowly if the capsule was not opened in the first instance. If fissures have extended to the cartilaginous investment, but not through it, as may occur,— the joint may remain unaffected until the suppuration extends along the fissures, the cartilage then yielding, gives way and the joint inflames; in such cases, the symptoms occur at a later period, suddenly and with great severity, so that a patient apparently in a satisfactory state may, in the course of a few hours, present the most severe symptoms.

Inflammation of the joint may occur subsequently, when the bone merely is injured, the capsule unopened and no fissures extend to the joint, on account of the spongy texture of the bone becoming necrosed as far as the previously described hæmorrhagic extravasation extends in the bone, and as the extension is generally limited by the cartilage of the joint on one or the other side—on the occurrence of inflammation of the bone in the whole extravasated portion, the cartilage gives way and the joint is broken into. We found in many cases such round sequestra in the head of the humerus, and already separated from the surrounding parts by granulations, but yet they subsequently excited suppuration of the joint and rendered resection requisite.

### 3. On the Diagnosis of Injuries of Joints.

From the preceding observations it may well be concluded that the injuries of joints are often very difficult to diagnose. Frequently the outer aspect shows that the joint is comminuted—the extensive swelling, and severe fever quickly developing itself, are enough for the experienced surgeon. The finger discovers the numerous splinters of bone, or the smooth surface of cartilage. Pressure causes thick synovia with blood to escape.

In other cases one or other of the symptoms fails, and the characteristic symptoms arise only at a later period. Hence all gun-shot wounds, near a joint, must be examined as closely as possible, we have often seen the most skilful surgeon err in his diagnosis.

It is well known that bullets may course semicircularly or further under the skin and again escape, or remain lodged. We have observed this also in the neighbourhood of the larger joints, as at the knee, elbow and shoulder, from the direction one would have concluded the joint to be injured, but it was not found subsequently to be so. Such cases are, however, seldom enough, and hence must not be at once concluded to be present, when one is unable to enter the joint or to feel splinters. Cases occur in which tendinous or muscular tissues slip over the wounded spot, and check both the escape of synovia, and the introduction of the finger, although a considerable injury of the joint may be present. A case of this kind will be related under 'Resection of the Knee'. Hence it is of the utmost importance to place the limb in that position in which it was, when wounded,—if the patient does not remember, it must be tried in various postures.

Distension of the capsule by serous or bloody fluid soon after the injury, is not a positive sign that the same is injured, as such may accompany the inflammatory state of the neighbourhood.

In cases where the joint is opened after the commencement of suppuration, deposits of pus form, as already described, between the cellular layers of the limb, and after opening these abscesses, if by pressure over the capsule a quantity of serous matter flows from the incision, this plainly declares that the joint has been injured.

## 4. On the Treatment of Injuries of Joints.

On account of the sad results usual in injuries of great joints when left to nature, the treatment must be very energetic and operative interference is generally necessary. The peculiar structure of the joint is the cause of this danger, hence, as the life of the patient is generally threatened under the most active antiphlogistic treatment, it is the duty of the surgeon either to remove the whole limb, or to simplify the wound by more or less complete resection. In certain cases it is sufficient to open the capsule freely, so as to allow the copious secretion to escape. Then, however, the healing is usually after long suffering, and ends with complete anchylosis, while by resection the cure is essentially hastened, and often gives a greater or less mobility of the limb.

In injuries of smaller joints, as of the fingers and toes, a cure may be expected without operation; as in many cases the bone is entirely destroyed, or otherwise the discharge of the cartilage and cure by anchylosis can be expected without much injury to the whole organism. Some trials by B. Langenbeck, however, have shown that even in such cases, when the preservation of motion is of great use, as in the finger-joints, a resection of the same may be carried out with good effect, if care is taken by early-employed passive motion to ensure a new joint. Unfortunately, however, in this kind of injury the tendons are usually involved in the inflammatory and suppurative process, and as they then slough away or adhere to the neighbouring parts, —the passive mobility of the joint is of no use to the patient.

Injuries of the hand and foot heal frequently, though after long suffering and with anchylosis, if they are suitably treated and the external circumstances are good, hence amputation seemed to us to be indicated only when extensive comminution of bone was present. Resections, however, we did not perform because, on the one hand, they are difficult to perform without much injury to the numerous vessels, nerves and tendons around the parts, and, on the other hand, because they seemed to promise no better result than those attainable by free opening of the capsules in the proper places.

The wounds of the great joints of the extremities, however, require almost always operative help, whether the bone be injured or the capsule merely opened by the bullet to a slight extent. By active antiphlogistic treatment and large doses of opium in the later stages, we have succeeded under favorable circumstances to cure certain such cases, though almost always with anchylosis. Gun-shot injury of the knee, elbow and shoulder, in which the bone is likewise comminuted, may be cured under specially favorable auspices without operation. The cases are however so rare, and the life of the patient is in such danger throughout, that only for special reasons can the surgeon justify himself in declining to operate.

Hence, in our opinion, in all gun-shot wounds of the four greater joints of the extremities, in which the bone is at the same time injured, the question is only, whether the limb shall be removed, or its preservation attempted by the performance of resection.

Conservative surgery exerted a prevailing influence in the Schleswig-Holstein campaigns, and in general we could be well satisfied with the results. In former wars, limbs have been almost always amputated for comminution of the joints by musket-balls; the resection of the shoulder-joint alone had been introduced by Larrey into military practice. According to the reports of the Academie Française, in 1848, during the revolution, very few resections of joints were performed, although the most celebrated French surgeons treated the patients. Pirogoff ("Rapport Medical d'un Voyage en Caucase," etc., Petersbourg, 1849), relates that in the war in Caucasus he performed in all 30 amputations or exarticulations of the upper arm for comminution of the elbow-joint or humerus; the results of his operations are not unfavourable, yet he seems to have amputated for every injury to bone, as, at the conclusion, he proposes the question, "Whether in fracture of the arm by fire-arms it is allowable to make an attempt to save the limb?" * I have already mentioned that in the last campaign we no longer considered even extensive comminution of the humerus as an indication for amputation; in 40 cases of comminution of the elbow-joint, by means of resection we have attempted to save life and limb; of these cases 6 died, in one the fore-arm mortified, one patient is still under care, the remaining 32 are cured, and all retain a more or less useful arm. The comparison of the results of operations of resection and amputation, at least in the upper extremity, proves that the first is by far the least severe and dangerous operation. As the most of these resections were performed under circumstances in which more than a third died of those amputated in the arm, this number must indeed be considered as favourable to conservative surgery.

It is hardly necessary to specify the further advantage of resection, the helplessness of a man without an arm, speaks sufficiently for the retention of even an impaired member.

In wounds of joints of the lower extremity the case is somewhat different, a good artificial or wooden leg is more useful than a limb that is bent, unnaturally moveable, or otherwise troublesome. A further question is as to which operation is the most dangerous for the patient, and this point I consider to be undecided. The attempts we made in resections of the knee and hip, though proving fatal, yet certainly justify and demand further attempts, —perhaps this question will be decided in the next great campaign.

The objections against resections, that the removal of the patients is not easy, must not be allowed to carry weight as regards wars in Europe, which are commonly between civilized nations, who neither mistreat nor murder the wounded prisoners. Those severely wounded should not be far removed, and if misfortune gives the enemy possession of the place, skilled and courageous surgeons should be left in charge, and allow themselves to be taken prisoners.

We have learnt from Danish surgeons, that they performed no resections, but in some cases attempted to preserve the limb by the extraction of all the loose fragments. This proceeding is, however very difficult and often impracticable in injuries of joints, from the tendinous attachments of muscles

* Pirogoff operated 45 times on the upper extremity, of which there were 9 exarticulations of the arm, 10 amputations of the arm in the upper third, 11 in the middle third, 7 amputations of the forearm, 8 amputations and exarticulations of the thumb or metacarpus. Of these 45 patients, 7 died.

and various bony prominences hindering a complete loosening and extraction of the splinters. By this means the evil consequences of joint-injuries are seldom averted, as the joint is not destroyed and the cure is only obtained by anchylosis. Should a complete removal of all the splinters be undertaken, an equally extensive division of soft parts would be required as for resection, so that it would be preferable to perform that operation after a regular method.

## 5. On the Injuries of individual Joints.

The decision of the question as to the preferability of amputation or resection for a wounded joint requires special rules. To show the results of our observations and the methods employed, it will be suitable to consider the joints singly.

### a. *On Injury of the Shoulder-joint.*

#### Diagnosis.

The diagnosis is often extraordinarily difficult even with considerable comminution of the bone; in wounds of the capsule alone, it may be impracticable to decide with certainty. The cause of this difficulty consists partly in the thickness, often very great, of the deltoid muscle, as it extends over the anterior, external and posterior surfaces. The swelling shortly following, and the circumstance that the limb was generally at the time of injury in an entirely different position than now when examined, and that thus the openings in the integument and deep parts do not correspond,— increases by much the difficulty of examination. If the bullet enters the axilla as frequently happens, and wounds the joint on its inner side, the examination is still more difficult, although the os humeri is here uncovered by muscles; as in such cases contusion of the axillary vessels must be present, great care must be taken not to abduct the arm strongly, as a rapidly fatal hæmorrhage might thus be caused.

To this must be added, that in general in injury of the shoulder-joint, the ordinary symptoms of a joint-injury are wanting. Even in considerable comminution of the bone, the pain at first, whether from touching or moving the limb may be but slight, and usually neither a collection of synovia nor its escape can be recognized; this evidently arises from the capsule not possessing any processes in which the secretions may collect, and as it is close-fitting, and if opened, is generally so, as far as its lowest limits, and as it is compressed by the deltoid on three sides,—the synovia escapes gradually and constantly, mixing itself with the other secretions without being able to render them peculiar.

Fortunately the shoulder-joint is that one of the larger joints, which suffers least from neglect. The inflammatory symptoms do not arise with so much severity, the supperation is more tardy and threatens the life of the patient less, than in injuries of the knee or elbow; probably because retention of pus is not so easy of occurrence. Secondary resections also appear to give an almost equally favorable result, as those performed primarily. The difficulty of the diagnosis was acknowledged by earlier authors, as by Larrey and Guthrie, and the latter indeed advises not to continue the examination too long nor too anxiously, as from this, more harm than good might arise.

### Course of Injuries of the Shoulder-joint without operative Interference.

Injuries of the shoulder-joint by a sabre-cut (according to Larrey) and also other simple injuries by a gun-shot, heal more readily than the wounds of other joints. Guthrie has already shown that comminution of the bones entering into the joint, may heal without resection by energetic antiphlogistic treatment and under favorable circumstances; he relates two cases, as of exceptional character. In the last campaign, we observed three such cases, in which the injury was not known at first,— later as the symptoms seemed favorable, they were left to nature. In most cases, however, the pus became so profuse and foul, that life was greatly endangered. Numerous sinuses formed in the arm, on the back or on the breast and required incisions, pyæmia generally occurred later, proceeding from the medullary canal, and death mostly followed from hæmorrhage—due to obstructive clots in the veins, so-called pyæmic.

If such cases are to be treated without a severe operation, yet successfully, —in the first place the strictest antiphlogistics, venesection, leeches and ice, must be used, every collection of pus be quickly opened, the wound be dilated with the bistoury and all loose splinters be as far as possible withdrawn. In every case the entire healing of the wound is much delayed, and only occurs on the completion of entire anchylosis, following the discharge of loose splinters, foreign bodies and the cartilaginous layers. The three cases above-mentioned are still under treatment (April, 1851. ), the anchylosis has begun, but necrosed bone and fistulæ remain, which render it doubtful whether an operation can be dispensed with.

### Injury of the Shoulder-Joint rendering Exarticulation of the Arm necessary.

Exarticulation, on account of injury of the joint, is but rarely indicated; namely, in the case where the bullet enters on the inner side, and simultaneously injures the axillary vessels and nerves,—the hæmorrhage or mortification of the limb may require this operation. Again, if the soft parts are torn away on the exterior of the limb by the heavier projectiles, and the joint is simultaneously much injured, exarticulation is to be performed by a flap from the inner side; if the loss of soft parts is less considerable, preservation may be attempted by means of resection, as the loss of substance replaces itself astonishingly, by gradual drawing together of the skin of the back during cicatrization.

If the joint is not injured, but the capsule laid bare widely, the case must be treated as a simple flesh wound with loss of substance. In such a case, where a great portion of the deltoid muscle was torn off by small shot discharged almost 'a bout portant', we could observe the capsule for the extent of more than a crown; after sloughing of the cellular tissue and cleansing of the wound, granulations gradually formed and healing followed without affection of the joint.

### On Resection of the Shoulder-joint.

The results of resection of this joint proved so favorable, healing being rapid, and frequently so much mobility remaining, that we do not hesitate to enunciate the principle, that in all injuries of the shoulder-joint complicated with injury of the bone by gun-shot, resection should be at once performed, if other complications are not present rendering exarticulation expedient. Of 8 cases left to nature, 3 alone did not prove fatal; and in these, after a six month's treatment, an operation may still be necessary. On the other hand, of 19 patients in which the shoulder-joint was resected, only 7 are dead; while 12 are healed, and preserve a more or less useful arm. Of these, some were healed in between two or three months' time; few complications occurred in the treatment, except some collections of matter requiring incisions, generally caused by sequestra, which were discharged at a still later period, necrosed. Complete anchylosis occurred in none, sufficient mobility remained in all; in many, however, the active mobility of the arm returned to so great a degree, that the patients could even perform heavy work. In the fatal cases, the resection was either made in the period of the highest inflammation, or the patient was among others, still less severely wounded, but suffering from pyæmia.

Larrey and Guthrie have given as a principle, that resection of the shoulder must be only performed when the head of the humerus alone is injured: that if the comminution extends to the medullary canal of the shaft,—or if fissures merely pass into the shaft,—the arm must be exarticulated, as after resection such severe symptoms are apt to occur that the patient, in the majority of cases, does not survive.

Our experience contradicts this principle most decidedly. The greater number of injuries for which we successfully resected the os humeri, also affected a greater or less portion of the shaft,—indeed in certain cases the portion resected was four or five inches long. The head is seldom alone affected, generally the borders of the epiphysis and diaphysis are struck, so that both are comminuted:—or the head is not struck, but merely the neck of the humerus, in such cases in young individuals the fissures terminate at the end of the shaft, and if the capsule is not injured at the same time, nature may be trusted to,—if it should be wounded, resection is indicated and should be performed immediately the joint becomes involved.

If the bullet strikes the shoulder-joint on the outer side, it may after comminution of the same, pass on, shatter the scapula, or enter the chest. If the last is the case and such severe injury to the lung is present that life is hopeless, of course no operation can be thought of. If the symptoms are of less amount, resection may be at once performed, as the loss of blood during the operation can but be useful.

If the glenoid cavity, or the anterior costa of the scapula are also comminuted, the loose splinters may be removed and the sharp points sawn or clipped off at the time of resection. Should fissures extend still further into the scapula, I believe further resection is not necessary, as the fissures heal more readily than in long bones, and at the worst a few gravitating abscesses form in the back.

We performed the greater number of the resections of the shoulder after Bernard Langenbeck's method: by means of a simple longitudinal incision on the anterior aspect of the joint, commencing at the edge of the acromion and following the course of the long tendon of the biceps downwards, for from two to four inches. (The chief point is to preserve this tendon, which has been usually divided by former operators. Remembering the anatomical relations of the tendons to the bone, this operation is easily performed on the corpse). Integument and deltoid muscle being now divided, the long tendon of the biceps is visible, its sheath is opened on the outer side, and the back of the knife being laid in the groove by the side of the tendon, the point is pushed up into the joint, always on the outer side of the tendon,—the cartilaginous surface of the head of the humerus is now visible. A blunt hook is now used to draw the tendon, luxated from its groove, to the inner side, and the edges of the wound are similarly held separate. The arm is now rotated inwards, so that the whole tuberculum majus becomes visible, and by carrying the knife semicircularly around and outside this, the tendinous insertions of the teres minor, supraspinatus and infraspinatus are divided. The arm is now rotated externally until the tuberculum minus is seen and at the same time the long tendon is carried outwards, and a semicircular incision around this tubercle similarly divides the tendon of the subscapularis muscle. These incisions being made so as to divide the capsule and the parts directly covering it in front, the head of the humerus may be easily forced out of the wound by drawing the long tendon inwards, carrying the lower end of the humerus backwards, the upper end forwards with respectively the right and left hands. The posterior part of the capsule is now divided, and the attachments of the pectoralis major, teres major and latissimus dorsi muscles are separated downwards as far as it is wished to remove the bone. The elbow being now carried upwards and backwards, the upper end of the humerus may be so easily brought out that a common amputating saw may be used to divide the bone. The arm is then brought to the side and the wound closed.

In this way the operation may be easily performed on the living, if the head of the humerus is grazed or split longitudinally, and not entirely separated from the shaft. The swelling of the soft parts, often very considerable, alone increases the difficulty of the first part of the operation, but rapidly goes down by the escape of serum and the loss of blood. If the head of the humerus is entirely separated from the shaft, or broken into many portions, the operation is much more difficult as the head cannot be rotated with the arm. In such a case, Dr. Franke carried a transverse incision two inches long outwards, from the upper end of the long incision, and separating this part of the deltoid from the acromion obtained sufficient room to be able to free and take out the comminuted head; this method was later followed a certain number of times with good success.

The chief difficulty, however, that of the head being separated from the shaft, requires either large forceps, or the sharp hooks invented by Langenbeck to fix the head by forcing them into its spongy portion, so that motion in all directions is effected at will. In such cases, the part to be divided is never well seen, and hence one must operate, so to say, blindly. The index finger of the left hand must now constantly accompany the knife, as by this means only the stretched ligamentous parts can be felt. For such divisions, and especially of the posterior part of the capsule, a long blunt-pointed scalpel, invented by Langenbeck, proves of great use.*

* All these instruments invented by Langenbeck, were supplied with the new army instruments, and are pictured in a Dissertation by Dr Petruschky of Berlin, "De resectione articulorum extremitatis superioris". Many preparations of bone, removed by Langenbeck in the first campaign, are also figured.

As soon as the separated pieces of bone are thus dissected out of the wound, the lower fragment-ends may be thrust out, and either the whole broken surface be sawed off or only the point, if the splintering extends far downwards.

As in Langenbeck's method the pus cannot well escape in the recumbent posture, Stromeyer devised another plan by which this want is corrected and the operation is equally easy. He makes a semi-circular incision, commencing at the posterior edge of the acromion, and carried downwards for three inches, its convexity outwards, along the posterior edge of the outer surface of the shoulder. The joint is thus reached from above and behind, it is widely opened with ease, and the humerus being raised forwards, the anterior parts of the capsule and the long tendon appear so freely, that they may be easily divided without injury of the long tendon,—for this the blunt-pointed bistoury is useful.

If the bullet has traversed the joint antero-posteriorly or the reverse, this method is not of particular use, as free escape of pus takes place behind, and therefore in these, far the most frequent cases, we employed Langenbeck's method which is the quickest in any instance.

The transverse separation of the muscle from the acromion by Stromeyer and Francke, did not seem to interfere with its usefulness, as its upper edge applied itself to and united with the articular surface of the scapula, it was thus fully attached and able to raise the arm, the healing was also quicker as the space to be filled by granulations was much diminished in size, by this application of the muscle to the glenoid fossa.

It appears that the long tendon of the biceps has been hitherto always divided by operators; we always tried to save it, because we thought it proper to save every muscular attachment possible, although the doing so renders the operation much more difficult. Possibly the tendon, naked and free in the wound always sloughed and escaped with the pus, we never observed such pieces, but it may have been so. That its preservation is not essential was shown in three cases, when the tendon had been torn across by the ball, yet on the cure being completed, the patient very soon obtained free and voluntary use of his arm.

Hæmorrhage during the operation is at first very free from the inflammatory enlargement of the smaller arteries, it soon ceases, however, if pressure by the finger be employed on the open mouth, and as some loss of blood is serviceable for the patient, it is not necessary to compress the subclavian. If the humerus is sawn off so low that the insertion of the latissimus and teres major must be separated, it is scarcely possible to avoid the posterior circumflex artery, which therefore towards the end of the operation spurts with violence. It must be ligatured.

After the operation we usually closed the wound by interrupted sutures, leaving a portion—or the posterior shot-aperture—open for the escape of pus; the arm was flexed and bandaged to the chest and body, and cold water dressing applied to the wound. Frequently in three or four days a part had healed by the first intention, and laudable suppuration set in, the cold dressings were then changed for warm ones, or merely charpie dipped in oil was applied, and exuberant granulations soon arose and filled the wound. The surface was then dressed with nitrate of silver lotion until the rapid cicatrization took place. If the pus was retained, or formed gravitating abscesses in the arm or back, these were opened or the wound dilated, and thus frequently small splinters could be withdrawn from the now-necrosed, sawn surface; we must never attempt to press the pus out of the wound at the time of dressing, but merely cleanse it, by trickling over, or syringing it with warm water.

Passive motion, very limited at first, gradually increasing in strength, was begun directly cicatrization commenced, and we were able to preserve considerable power of motion in all the cases that recovered. Gradually the muscles of the patient began to develop their activity, and in many cases so quickly that the hand could be raised to the mouth before the wound was quite cicatrized.

Of 19 resections of the shoulder, 7 only were fatal, most of them from pyæmia. In 5 of them before death active hæmorrhage set in, caused by obstruction in the veins and was not to be stopped by ligature of the artery: as shown by one case, in which the axillary and subclavian vessels were successively tied In such cases, either foul suppuration in the medullary canal, or phlebitis of the axillary vein, or both were always present.

The length of interval between the injury and operation seemed to exert an influence on the result. Of 6 performed in the first twenty-four hours, but 2 were fatal. In the stage of commencing suppuration, hence in that of the highest inflammation, on the third or fourth day, 3 resections were performed, of these 2 proved fatal. Secondary operations, that is, after the full occurrence of suppuration,were effected 10 times, with fatal result in 3 cases; which is somewhat more favorable than in primary resection. Hence, I believe, it would be better to resect earlier, directly after the reception of the injury, or at least within the first twenty-four hours. If the injury is first discovered on the third or fourth day, or one is prevented from operating sooner, than it is advisable to wait till suppuration is fully set in, and until that time, to moderate the inflammation by strong antiphlogistics.

It is curious that the operation on the left side seems to give less favorable results than on the right, 6 of 12 died of those resected on the left; 1 out of 7 of those resected in the shoulder on the right side. A similar proportion held good in resection of the elbow, in whom, of those operated upon on the left 4 in 19, on the right, 2 in 20, resections proved fatal. From this, the fatality attending operations on the left arm to that on the right, is as three to one; but of course further observations are required to enable conclusions to be deducted.

To enable the reader to form his own judgment on the principles here laid down, there follows here short histories of such cases as I have been able to collect, and at the end, a tabular appendix of them. [Of the nineteen cases eight are translated.]

Case I.—Shot in the left Shoulder-joint, with Comminution of the Head of the Humerus. Resection on the 17th day. Recovery.—The Prussian Grenadier, Carl O——, aged 24, was wounded, 24th April, 1848, at Schleswig, in the left shoulder by a bullet, the head of the humerus being shattered. The bullet had entered near the coracoid process, could not be found and first appeared superficially two years later, near the inferior angle of the scapula. The general state was at first good, the severe pain being relieved by ice.

Gradually, however, becoming worse, the pain and also the suppurative discharge increased. On a more careful examination the joint was found involved. Dr. Langenbeck now, the 10th May, performed resection, with preservation of the tendon of the biceps.

The bullet had traversed the humerus, close below the joint, and fissured it freely upwards and downwards, so that 4½ inches of bone required removal by the saw. A small semicircular piece had been struck out from the glenoid fossa, and was forced against the lower surface of the scapula, which was evidently grazed and splintered, as matter formed here at a later period, and with it were discharged sequestra and pieces of clothes. Healing took place favorably, and the patient could go out by the end of July. On August 28th, the wound was closed, excepting a small fistula, which led to behind the scapula. At this time a more careful examination showed that the resected portion of bone was replaced by a fibrous mass, which felt tolerably firm, and bound the upper end of the humerus with the glenoid cavity; as this substance at a later period gradually hardened, its partial ossification was to be expected. The active mobility allowed the nose to be taken hold of by the fingers of the left hand. After the wound was fully healed, passive motion was regularly employed and the patient urged to make use of the limb. In the beginning of 1850, the upper end of the humerus had regenerated itself so much, that the resected arm was only one inch shorter than the other. It could be moved by the patient in all directions, and a chair could easily be lifted by it from the ground. The arm was as well nourished [robust] as the sound one, but the roundness of the shoulder was wanting, and it appeared as a luxation into the axilla. Also on each movement, the arm escaped somewhat from the glenoid cavity, backwards or forwards, partly from the ligaments being looser, and partly from the biceps tendon having lost its groove in the head of the humerus. In the summer, 1850, the bullet came to the surface, and was cut out.

Case X.—Grazing of the Os Humeri, and Comminution of the Spina Scapulæ. Resection on the 4th day. Hæmorrhage, Pyæmia, Death.—The Schleswig-Holstein foot soldier, B——, was shot in the left shoulder, at Fridericia, July 6th, 1849. He was brought to Hadersleben at once. July 10th, the arm and shoulder were greatly swollen and painful, and serous, thin pus flowed copiously from the wound. The bullet had entered at the posterior and outer border of the shoulder, an inch below the acromion; on examination, the finger reached the joint, and felt the os humeri bared on its posterior aspect; some splinters lay in the shot-track, the bullet could not be discovered. Dr. Goetze at once performed resection after Langenbeck's method; on bringing the head of the humerus out of the wound, it was seen that the bullet had grooved it posteriorly and also its anatomical neck, for an inch and a-half, the groove being three-quarters of an inch wide and one quarter of an inch deep. An inch and a-half was removed from the humerus, and the arm returned to its position. Fissures of the scapula were perceptible on examination, especially about the superior notch. Some loose splinters were removed, the bullet was not discovered. At first proceeding well, the patient later became feverish, and, July 17th, hæmorrhage occurred about midday, after the wound had been dressed. Six ounces of blood were lost. As the source of this bleeding could not be seen, the wound was dilated, some sequestra were removed, a few small vessels were twisted and the wound filled with charpie. The hæmorrhage on the following day returned, oozing from the granulations in different places and of no decided arterial colour. It was justly feared that the bleeding was caused by stoppage in the veins. Some bleeding places were again twisted, the wound plugged by charpie and ice applied to it. The patient was now very weak, the pulse small and quick, but somewhat hard. Slight rigors occurred. On the evening fever became exacerbated, the wound dry, and watery pus oozed from it; the periosteum began to separate from the humerus, and the whole arm became œdematous. On the 22nd July, there were several rigors and delirium, and, on the 23rd, death ensued.

On the autopsy, there were loose coagula in the axillary and subclavian veins, the liver full of metastatic abscesses, the bullet lay flattened behind the neck of the scapula, from which fissures proceeded in all directions, so that by them a portion of the spine of the scapula and the coracoid process were fully separated. Foul pus had collected in the fossa supraspinata.

Case XI.—Grazing of the Os Humeri. Partial Necrosis of the same. Resection on the 35th day. Cure.—Hans. L——, a Schleswig-Holstein rifleman was shot at Fridericia through the right shoulder on the 6th of July, 1849, and carried first into Veile, and afterwards, on the retreat of our troops out of Jutland, into the hospital at Hadersleben. On the 20th of July, the day of his arrival, the following was the result of the examination. The ball had penetrated at the middle of the outer border of the shoulder-blade, and passed outwards again from the coracoid process. The shoulder-blade was not injured, but we found on examination with the finger through the orifice of exit, an extensive injury of the humerus; its continuity however was not destroyed, for the head of the bone rotated with it. The inflammatory swelling of the arm and shoulder was not very great; the fore-arm and hand were œdematous. The suppuration was favorable, but rather profuse, the region of the shoulder was not very sensitive, either on voluntary or passive motion. The general condition of the patient, which before was tolerably good, had been rendered worse by the journey: violent fever, diarrhœa, and loss of appetite supervened, and he complained of great weakness. Though there was no doubt that the joint was injured, still under these circumstances we decided not to operate. The diarrhœa soon yielded to opium and nitre, and we proceeded to give tonics. With acids, wine, beef tea, &c., the patient made daily progress, although the suppuration rather increased than diminished. On the 10th of August, after a fresh examination of the wound, resection was determined on, and immediately performed by Dr. Francke. He made, according to Dr. Stromeyer's rule, a five-inch long semicircular incision through the skin and deltoid-muscle, which, beginning at the posterior edge of the acromion, was continued backwards and forwards. After the capsule was laid open from behind and from above, it was easy to cut through the remaining part of it, together with the muscular attachments, and thus to free the long tendon of the biceps from its groove without injuring it.

The head of the bone was now forced out through the wound, and although the inner part of the shaft was shattered still farther downwards, only a piece three inches long of the upper end of the humerus was sawn off. After the sharp edges were clipped off, and some loose splinters removed, the humerus was restored to its proper position, the upper horizontal part of the incision was drawn together by interrupted sutures, and the arm, bent at a right angle was fastened to the body. The piece of bone sawn off exhibited the following conditions. The inner and under part of the head, and a large piece of about two cubic inches of the inner side of the neck and shaft, with its superincumbent tuberculum minus, had been crushed and torn away by the bullet. In the middle of the portion removed, lay still a large piece of spongy bony substance, which had plainly become necrosed through the injury and inflammation; in some parts this still adhered to the remaining bone, but in most places it had been thrown off by the granulations which had arisen out of the uninjured bone, shewing that the sequestrum would in a short time have been entirely separated; some pieces of cloth stuck firmly to it. A transverse fissure running from the injured spot under the tuberculum majus along towards the outer and posterior part, was already filled and healed by fresh bony substance. The articular surfaces were for the most part bared of cartilage and covered by granulations; and after maceration it appeared that new masses of bone had already formed under this thin covering. It was plain to us after this discovery, that after separation and rejection of the sequestrum, a natural cure with anchylosis at the shoulder-joint would have been possible, if the patient had been able to bear long enough the great loss of secretions. The reaction consequent on the operation was trifling: the pain soon ceased after a dose of half a grain of morphia, and the patient had a quiet night. On the third day, the bandage was removed, healthy suppuration set in, and granulations of good character everywhere soon appeared. Pus flowed principally through the upper transverse part of the incision, and the sutures which threatened to cut through had to be taken away, as the fibres of the deltoid muscle hindered the flow of matter out of the long incision, and this speedily healed. As soon as the whole wound had covered itself with granulations, and cicatrization commenced, the lower border of the transverse incision drew itself more and more into the wound, and united itself by degrees with the articular surface of the shoulder-blade, forming here a much depressed cicatrix under the acromion. Latterly the arm was fastened by a bandage, which, like Desault's, raised it towards the shoulder, and thus the patient left his bed in the middle of September. His strength increased from day to day, and towards the end of October, he was dismissed from the hospital cured. The passive mobility of the shoulder-joint was already very great, but only the first signs of active mobility shewed themselves. In the beginning of 1850, he was already able to abduct his arm to almost a right angle from the thorax, and to carry heavy weights, the fore-arm and hand having their full strength. For this, that portion of the deltoid muscle was of especial service, which had been severed by the upper transverse incision, and had afterwards united with the articular surface of the shoulder-blade. After L—— had returned to his former occupation of a labourer, I heard from an eye-witness in 1850, that he was even able to thresh, a work certainly requiring no little use of the shoulder-joint.

Case XV.—Splintering off of the upper end of the Humerus. Resection on the 1st day. Healing.—The Schleswig-Holstein rifleman, John L——, was shot on the 12th of Sept., at Missunde, through the right shoulder, and carried into the hospital at Rendsburg. The bullet had penetrated the front of the shoulder, two inches under the acromion, under the groove for the biceps, and had passed out at the back about three inches under the acromion. Examination with the finger showed comminution of the humerus in its surgical neck, and I performed, therefore, on the following evening, resection of the head of the humerus by making a three and a-half-inch long incision outwards from the acromion, transversely through the aperture of entrance. The long tendon of the biceps had been divided by the bullet and there was no need, therefore, to take care of it. The circumstance that the head of the bone had been entirely separated from the shaft, and, consequently, did not follow the motions of the arm, caused some difficulty; but Langenbeck's sharp hooks fixed in the head of the bone, rendered good service. After the capsule, together with the muscular attachments, had been divided on all sides, the separated part of the bone was easy to remove. This was of considerable length. From the spot where the ball had struck, extended a slanting fissure three inches long outwards, and a fissure two inches long inwards. Thus the severed piece of bone assumed the shape of a molar tooth having two unequal fangs; its greatest length being five inches. After the two sharp points of the under fragment were clipped off by the bone-forceps and some loose splinters removed, the wound was partly brought together by sutures, covered with oil-dressing and charpie, and then the arm was fastened to the body in a semiflexed position by two triangular cloths. A bladder of ice was laid on the shoulder. On the third day, suppuration had set in, the sutures were removed, and the wound carefully washed out with warm water. At first, as long as sloughed tissue escaped, the pus had a bad appearance, but, by the 20th of September, it became normal, and the quantity diminished daily; the opening of the exit of the shot-canal, on the back part of the shoulder, was very useful for the discharge of matter. The fever was trifling, the general condition satisfactory. On the 28th of Sept., as I paid my morning visit, the surgeon in attendance told me that an abscess had formed on the under and outer border of the deltoid muscle, which would soon require incision. Fluctuation certainly was felt here, but I observed that the surgeon was trying to empty the wound of the accumulated secretion by pressure from below, and thus, in all probability, had brought about the abscess artificially. I begged him, therefore, to refrain from pressure for some days, in order to allow the pus to accumulate there, and only to syringe the wound a little with warm water. The consequence was, that there was no collection of pus, and, on the 1st of Oct., every trace of fluctuation had disappeared. Some small splinters came away at times, and then the whole wound quickly filled with exuberant granulations, necessitating the application of a weak solution of caustic, under the influence of which the hollow soon lessened, and the edges of the wound exhibited margins of cicatrization. On the 10th of Dec. the patient first left his bed. The wound was almost entirely cicatrized, the general condition excellent. He was then moved into the convalescent hospital at Jevenstedt, where I saw him again on the 17th of January, 1851. The passive mobility of the shoulder-joint was very great. The humerus was two inches apart from the acromion. He was already capable of slight active motion, and became daily more so. The hand had its perfect strength, the elbow-joint was flexible, and it was only incapable of perfect extension because it had so long remained bound in a bent position.

By the beginning of March the active mobility of the arm was considerably recovered. The patient was able to abduct the arm to an angle of 40 degrees from the thorax, in doing which the posterior portions of the deltoid muscle were of especial service. Even a slight rotation of the arm in the shoulder-joint was already possible. The flexion of the fore-arm was powerful ; a weight of 4lbs. could be lifted to the height of the left shoulder. On the upper end of the humerus formation of callus was traceable; but it was still separate about two inches from the glenoid cavity of the scapula.

Case XVI.—Shattering of the left Humerus. Resection on the 1st day. Death by Pyæmia.—The Schleswig-Holstein foot-soldier, Sch——, was shot on the 4th of Oct, 1850, at the storming of Friedrichstadt, through the left shoulder, and brought into the hospital at Delve. On account of the head of the humerus having been shattered, Dr. Herrich of Regensburg, undertook its resection on the following day, after Langenbeck's method. The ball had passed in on the groove of the biceps close under the acromion, and issued on the back part of the shoulder, had torn the long tendon of the biceps, and comminuted the head of the humerus. The fragments were removed, and the fractured ends of the humerus were sawn off. Altogether a piece of two and a-half inches long was taken away. At first, things went on excellently; the wound quickly granulated, discharged good pus, and the general condition was very satisfactory. Subsequently, gastric affections set in, the pulse became quicker, the tongue coated, the appetite was lost; and on the 20th of Oct. the patient suffered a severe rigor. This soon recurred daily: profuse perspiration followed, the tongue became parched, the wound yielded slight, fœtid matter. Delirium and death supervened on the 1st of November. On the autopsy, pyæmic abscesses were found in the lungs.

Case XVII.—Shattering of the Head of the right Humerus. Resection on the 1st day. Recovery.—The Schleswig-Holstein musketeer, Detlef K——, received on the 4th day of October, 1850, a bullet through the right shoulder, and was carried into the hospital at Rensburg. The bullet had passed in, close under the acromion, on the long tendon of the biceps, tearing through the same, piercing the head of the humerus, and breaking it into five great fragments, had issued on the anterior edge of the scapula. The day following, Dr. Dohrn, performed resection of the head of the humerus; he made, from the orifice of entrance, a three-inch long incision downwards, and one, two inches long transversely through the deltoid muscle outwards. The fragments of bone were now easily removed; the fractured part of the humerus rendered even with the bone-forceps, and the wound closed by sutures; four inches in length of the humerus were removed. The transverse incision healed by first intention; the remaining part of the wound very quickly granulated, and nothing wrong supervened except a trifling gravitation of pus on the fossa supraspinata, which required opening with the scalpel. Towards the end of November, the wound was almost cicatrized, the patient left his bed and began to move his arm, and soon acquired great mobility of it. On the 16th of January, 1851, the wound was perfectly cicatrized, the arm admitted of passive motion in every direction, without pain, and the active motion was then considerable. The patient was able to abduct the arm to an angle of 50 degrees from the body; and, with some trouble, could lift a weight of 3lbs. to the height of the left shoulder. Even a slight active rotation of the arm was already possible.

Case XVIII.—Shattering of the Head of the right Humerus. Resection on the 1st day. Recovery.—The Schleswig-Holstein musketeer John H——, was shot on the 4th of October, 1850, before Friedrichstadt, through the right shoulder-joint, and brought into the hospital at Rendsburg. Examination with the finger showed that the head of the humerus was pierced and comminuted: Dr. Francke, therefore resected on the following day, in the same manner as in the former case. The head of the humerus had been pierced by the ball antero-posteriorly; its outer part with the tuberculum majus was split into several fragments; the inner and larger part of the head was still continuous with the shaft, but was penetrated through its length by two great fissures. The splinters were removed, and the articular ends of the humerus sawn off at the surgical neck. This case also proceeded with no particular phœnomena: excepting that the transverse incision did not heal by first intention but by granulation, on which account the cicatrix was deeply drawn in under the acromion. and united there with the articular surface of the scapula. Towards the end of November, the patient left his bed, and commenced methodical exercise of his arm, but not so zealously as the former patient, and on this account he did not so far recover the active motion of the arm, as did the other one. On the 18th January, 1851, the passive motion of the arm was equal, but it could not be abducted to more than a right angle from the thorax, because then the cicatrix of the transverse incision became stretched, and the end of the humerus came into contact with the acromion.

Case XIX.—Shattering of the Head of the left Humerus. Resection on the 14th day. Hæmorrhage. Death by Pyæmia.—The Schleswig-Holstein musketeer, Ernst L——, received on the 4th day of October, 1850, before Friedrichstadt, a bullet through the left shoulder, and was carried into the hospital at Rendsburg. On his arrival, the shoulder was already considerably swollen, and a thorough examination was rendered difficult. The bullet had entered behind, between the axilla and the spina scapulæ, and had passed out between the axilla and the acromion. Through the front orifice, the finger reached the head of the humerus, which appeared still to be covered by the capsule; in the hinder orifice some splinters were felt. Notwi.hstanding constant application of ice, the swelling went on increasing some days, and a considerable effusion of blood manifested itself on the inner side of the upper arm, as far as the elbow, which was very tender. Still the general condition of the patient was little disturbed, the fever was slight, but increased a little on the 16th October, when the suppuration till then slight, became more profuse and fœtid, and on this day, a quantity of fluid, partly coagulated blood escaped, and was followed by visible decrease of the swelling of the arm and shoulder. The application of ice was discontinued, and the wounds were dressed with oiled charpie. On the 18th October, the arm again became swelled to a greater degree; fluctuation showed itself on the under part of the deltoid muscle; by means of a large incision a great quantity of fœtid pus was let out, and the shattering of the head of the humerus was distinctly felt by the finger. Resection of the same was now resolved on and performed by Dr. Thiersch, Prosector, of Munich. He divided the soft parts by a curved incision, extending from the anterior edge of the acromion to the posterior bullet-wound, opened the articular capsule, and freed the head of the humerus, which was split into three great and several small fragments. The outer and largest, extending downwards to a point, was altogether five inches long: on the inner side, the lower fractured part of the humerus extended with a sharp point nearly to the head of the bone; this sharp point was sawn off, so that there was only a loss of three inches of bony substance. During the operation the posterior circumflex artery spirted out very strongly, and it was difficult to close it by ligature. After the wound was in part united by sutures, the flexed arm was bandaged to the body. Rather high fever supervened on the operation, the patient complained several times of shivering fits, and on the 22nd October was seized by a violent rigor, followed by heat and perspiration. On the evening of this day there was considerable hæmorrhage, which, however, ceased spontaneously on removing the bandage and the sutures. The wound was again filled with lint, and muriatic acid, in a decoction of althæa was prescribed. On the following day the suppuration became profuse and fœtid, and the patient very feverish and weak, and his skin jaundiced. On the 24th of October, a rigor again occurred, recurring twice on each following day. Calomel and opium were given without result. On the 26th October, profuse venous bleeding again occurred, which ceased in like manner on removal of the dressings. The patient became weaker and weaker. On the 17th October, fresh bleeding came on and in a few hours he died.

On dissection,—the cellular tissue in the whole region of the shoulder-joint was found infiltrated with fœtid pus, and the upper end of the humerus bared of periosteum. The axillary vein for a length of three inches felt hard and corded; its coats were injected and thickened, and on incision we found a firm fibrinous coagulum adherent to the inner coat, and entirely stopping up its volume. We could not discover from what vessels the hæmorrhage had proceeded: the lungs, liver, and spleen were anæmic; but in none of these organs were pyæmic abscesses discoverable.

### b. *On Injuries of the Elbow-joint.*

Shot-wounds of the elbow-joint are in general more dangerous than those of the shoulder-joint; this is also agreeable to the old opinion. Thus Larrey writes in his "Clinical Surgery," "The rending of the fibrous and nervous structures of this joint is serious, and always demands amputation of the limb. I know no instance of a cure where this limb was seriously injured by a ball.* So also remarks Guthrie, in his "Gun-shot wounds of the Extremities," "Wounds of the elbow-joint by balls, even when only one of the bones forming it is injured, seldom result favorably: commonly they indicate at a later period the necessity of amputation, on account of destruction of the cartilage, after the hope of anchylosis of the joint has proved vain. By the great number of failures to save the limb under these circumstances, I am thoroughly convinced that such cases are only rare exceptions in military practice." This excellent writer already recommends resection of this joint, instead of amputation of the arm in certain particular cases and under favorable conditions, and yet he himself never undertook such an operation; and hence we may conclude that amputation of the arm was commonly performed in former wars for the severer injuries of this joint. We have acquired, like

* However, the father of French Surgery, Ambrose Parē, has recorded two excellent cases of healing after wounds of this kind

experience in our campaigns of the danger of these wounds, and yet there were but few cases where we deemed amputation necessary. There is no doubt that the honour of introducing resection of the elbow-joint into military practice, belongs to the Schleswig-Holstein military surgeons; namely, to the Surgeons-in-Chief, B, Langenbeck and L. Stromeyer; for according to the scanty records of these matters in the battles of 1848 and 1849, this operation was never performed for gun-shot wounds, either in Paris, Italy, Baden, or Hungary.

### Different Kinds of this Injury.

On account of the many bony processes which project about the synovial capsule of the elbow-joint on all sides, and very much protect its position, an injury by a bullet of the articular capsule alone is very rare. Commonly a greater or less shattering of the bones attends it, and owing to the complicated structure of the joint, these wounds exhibit an extraordinary variety.

Sometimes only one or another portion of bone is contused, and fissures extend from the wounded spot into the joint; in other cases, one of the bony processes,—for example, the inner or outer condyle of the humerus, the olecranon, or the coronoid process of the ulna,—is severed by the bullet, and at the same time the articular capsule is opened. In most cases, however, there is considerable comminution, affecting either one or more of the bones. The perfect ossification of the epiphysis with the diaphysis is earlier with the bones of the elbow-joint than with those of the shoulder and knee. On this account, in cases of comminution which affect the diaphysis in the neighbourhood of the elbow-joint, the fissures extend themselves the more readily into the same, and thus it often happens that, where the humerus or the ulna has been pierced by a ball some inches distant from the joint, the joint participates in the injury sooner or later.

### Diagnosis of Gun-shot Wounds of the Elbow-joint.

The diagnosis of these injuries is seldom difficult, on account of the few soft parts covering the joint, and from superficial examination being easy. If a ball has passed straight into the joint, injuring the bones, it is in most cases easy to reach the splinters or the opened joint with the finger. The diagnosis is only rendered difficult in those cases in which the ball pierces the skin at some distance from the joint, and only injures the joint itself after first passing a considerable way through the soft parts. Thus, for example, we have seen cases, in which the ball, after piercing the skin in the lower part of the fore-arm, has passed along the ulna, separated its coronoid process, injured the joint, and lodged in the upper arm. In such a case, of course, it is impossible to follow the track of the ball with the finger, but the violent inflammatory symptoms rapidly following the injury, and which appear in the joint itself, usually enable us to decide with certainty that the same has been injured.

### Course of Wounds of the Elbow-joint without operative Interference.

Considerable serous infiltration of the surrounding soft parts commonly follows very quickly on the comminution of the elbow-joint by a ball; and in cases of somewhat extended splintering we are often able, within some hours of the wound being received to diagnose injury of the joint, by the outward appearance of the limb, and without any closer examination of the wound. In the worst cases of this kind, after a long transport, or with injudicious treatment, infiltration often sets in very rapidly along the large vessels as far as the axilla, and often becomes so considerable that blebs of mortification form themselves on the extremely distended skin. Such are very bad cases, and a fatal termination is only to be avoided by timely operative interference, for if once suppuration follows the inflammation, purulent infiltration is rapidly developed; that is to say, the inflammation of the lymphatic vessels and small veins which we have already described; and then an operation only hastens a fatal termination.

In some severe cases, however, and with active antiphlogistic treatment, these symptoms do not reach such a point as to threaten the life of the patient. The infiltration does not extend so considerably, the inflammation may be less severe; but no sooner has the latter extended itself over the whole joint than violent fever sets in, the patient suffers excessive pain, and cannot endure the slightest touch or movement of the joint. This continues to swell more and more, and because there is seldom space enough for the escape of pus through the wound, it forces its way, after a time, through the articular capsule in those parts were less resistance is offered, takes its course upwards and downwards between the intermuscular and subcutaneous cellular tissues, and forms the so-called gravitating abscesses on the fore and upper arm. The cartilage separates itself simultaneously from the articular surface: and the denuded surface of the bone offers an extensive space for the action of the fœtid matter. The whole arm becomes full of sinuses: the bones participate in the sanious degeneration, and death follows with symptoms of pyæmia, unless it is decided at the right time to lessen the danger by means of an operation; a determination often arrived at too late, for as soon as shivering fits have set in, operative interference in general only hastens death. In the most favorable cases, indeed, the patient may recover without an operation, but commonly only after long suffering, after many incisions have been made, and the necrosed pieces of bone have been removed by degrees. In these cases a perfect anchylosis of the joint cannot be avoided.

The following cases show with what difficulty, even unimportant injuries of the elbow-joint, are healed without an operation:—

Case I.—Grazing of the Humerus above the Elbow-joint. Pyæmia Inflammation of the Joint. Opening of the same. Abscess. Healing.—The Schleswig-Holstein. private F. W.——, was shot on the 6th of July, 1849, before Fridericia through the back part of the left upper arm, and carried into an hospital at Hadersleben. The ball had penetrated half an inch above the internal condyle of the humerus, had grazed the humerus close above the olecranon, and bared it to a small extent of the periosteum, and had then passed out above the external condyle. Soon after his reception into the hospital, considerable inflammation of the wound came on, for which we used at first, leeches and cold applications, and afterwards warm fomentations. Copious suppuration now ensued, and as the secretion had no channel of escape, the orifice of exit was dilated. Good granulations now showed themselves, and the wound appeared about to heal, when high fever again supervened, and, suddenly, on the 20th of July, a severe rigor seized the patient, recurring at about the same time on the following day. Jaundice followed, and the wounds had a flabby, bad appearance. That this was a symptom of pyæmia could not be doubted, and indeed, several patients died of that disease at the same time, in the same hospital. A grain of calomel was now administered every two hours, the wounds being fomented with infusion of chamomile. Severe diarrhœa having set in, large doses of quinine were given instead of calomel. On the 28th of July the rigors ceased, the jaundice disappeared gradually, the patient was generally better, and by the 1st of August the wounds again looked well. On the 5th of August the elbow-joint became swollen and very painful, especially when examined. In the contusion the humerus had probably received a fissure which extended into the joint, and now first carried inflammation and suppuration into it. In spite of frequent local abstraction of blood, friction with strong mercurial ointment, and application of poultices, the inflammation and fever steadily increased and the general appearance of the patient became worse; the pain depriving him of sleep. On the 22nd of August an incision was made into the joint on the outside of the olecranon to facilitate the escape of fœtid matter. Upon this, the swelling of the joint subsided, though the general condition of the patient was still critical. By the middle of this month a metastatic abscess had formed on the os sacrum, which was immediately opened. As usual, a second then formed near the same place, and was also opened. In the beginning of September, a sore from recumbency formed on the internal condyle of the humerus, and immediately the whole upper arm in the region of the shoulder swelled to a great extent. A large abscess formed here, on opening which about twelve ounces of matter escaped. The patient's condition became better only towards the end of November, the various deposits of pus having been emptied by incisions. At the end of December, all wounds were healed, and the patient left the hospital with perfect anchylosis of the elbow-joint.

CASE II.—Splintering of the Olecranon and of the Head of the Radius. Extraction of the Splinters. Copious Collections of Pus. Doubtful Termination.—The Schleswig-Holstein private, H. T., was shot at Idstedt, on the 25th of July, 1850, through the right elbow-joint, and carried into the Princes'-palace at Schleswig. The ball had passed in at the point of the olecranon, shattering it and the head of the radius, and escaping on the inner side of the latter. There was no time for resection, so we only removed the loose splinters, and placed ice on the region of the shoulder. The inflammation of the soft parts being unimportant, and the pus escaping easily, we resolved to leave this case to nature. Pus formed at times on opposite parts of the fore and upper arm; the places were opened and loose splinters extracted. In the middle of August suppuration became very profuse, and the patient very weak. At this time, on account of the frequent incisions and the retraction of the undermined skin, there was a great loss of substance on the outer surface of the elbow-joint. Under strengthening treatment the patient became better towards the end of August; the numerous wounds filled with good granulations, and on the 10th of September all was looking well, when our surgeons left Schleswig, and he came into the hands of Danish medical men. I do not know how the case ended, as he did not return with the other wounded.

We shall see, as we proceed, that in many cases where an operation was not thought necessary at first, the progress of the case and the supervening symptoms rendered it indispensable.

Sabre-wounds of the elbow-joint often heal, according to Larrey, Guthrie, and Stromeyer, even when the bones are considerably injured; in such cases it is necessary, however, to extract the splinters of bone without delay, and anchylosis of the joint commonly follows. Larrey recommends also a very energetic antiphlogistic treatment, and especially general and local abstraction of blood. In our campaigns, I never saw a case of injury of the elbow-joint from a sabre-wound.

## On the Resection of the Elbow-joint

The foregoing account of the course which shot-wounds of the elbow-joint usually take, prove it, we think, the duty of the surgeon, in all cases of injury of this kind, to help nature by an operation. The great number of cures effected by means of resection of this joint, prove its superiority to amputation, independently of its evident advantage to the patient. Cases of course, occur in which amputation must be preferred, for example, when the ball has not only injured the joint, but also the vessels on its inner side, and when, in consequence, considerable bloody infiltration of the fore-arm follows, or subsequent arterial hæmorrhage necessitates the ligature of the brachial artery. In either case mortification of the fore-arm is to be dreaded,—in the latter, because the collateral branches behind the joint are generally severed by the resection, on which account immediate amputation of the fore-arm is advisable.

In cases also of shattering of the joint by round-shot, if the bones are extensively crushed, or the soft parts seriously injured or torn away, there is no hope of saving the limb by resection, and amputation is, therefore, to be performed.

We have tried resection of the joint in the remaining cases, and, as already mentioned, with the happiest results. On comparing, statistically, the results of resection of the elbow-joint with those of amputation of the arm, it can no longer be doubted which of the two is preferable. Out of 54 cases of amputation of the arm, 19 died; out of 40 resected in the elbow-joint only 6 died, and, therefore, cases of amputation of the arm for simple shattering of the elbow-joint by bullets, without other complications, became more and more rare amongst us. There were 6 such in 1848, 3 in 1849; in 1850, not 1, amputation of the arm on account of such injuries.

The healing of the wound after resection of the elbow-joint is indeed usually more tedious than after amputation, but then the patient retains his limb. Even in cases where the injury is trifling, and there is hope of its healing in time without operative interference, resection is still needed as soon as suppuration of the joint sets in; and not only are the sufferings of the patient considerably shortened, but many bad and dangerous consequences are avoided. At least, there is hope of obtaining power of movement in the elbow, while its anchylosis is hardly avoidable if the cure is left to nature.

For this last reason, in cases where the bones are only injured to a small extent, we ought not to be content with a partial resection of the wounded parts, with division of the articular capsule to a partial extent, for, in my opinion it is, indeed the extensive severing of the ligamentous apparatus of the joint which deprives the wound of its danger. The less there is removed from the joint-ends of the bones the greater is the probability of anchylosis. A partial resection was twice tried, in the last campaign, and although both cases terminated favorably, yet their course showed that such an operation is in no respect preferable to the entire division of the articular capsule, perfect anchylosis occurring in both cases, and the process of healing being far more tedious than in resection of the entire joint.

Case I.—Splintering off of the external Condyle of the Humerus, and Contusion of the Head of the Radiu , with Injury of the Articular Capsule. Extraction of a Splinter and Dilatatio of the Wound of the Joint on the following day. Tedious Healing and Anchylosis.—On the 12th September, 1850, a Schleswig-Holstein serjeant, Hans N.—— was wounded at Missunde in the right shoulder, and brought to Rendsburg. The ball had entered at the bend of the elbow in front of the external condyle of the humerus, and had passed out close behind it. As on the following day there was every appearance of a violent inflammation of the joint, an incision was made by Dr. Stromeyer on the outside of the joint, carried over the external condyle of the humerus, and the head of the radius, and the severed condyle were extracted. The articular capsule was torn close under the same, leaving the head of the radius exposed to sight. On rotation of the fore-arm it was seen that this had been hit and contused in one spot by the ball.

In this case Dr. S. wished, for the sake of comparison, to remove as little as possible of the bones, and contented himself with enlarging the wound of the articular capsule by the scalpel. As, after a resection, the arm was laid on a splint, and covered with an ice-bladder. Moderate, favorable suppuration began on the 16th of September, and the wound quickly filled with granulations. On the 24th of September, a small fragment of bone came away with the pus and a raw surface of bone in the depth of the wound was ascertained by the finger. Towards the middle of October the joint became very painful and swollen considerably on its back part, from feverish involuntary movements; on the inside of the olecranon the skin became red; fluctuation was soon perceptible, and, on the 18th of October, an incision in this part released a large quantity of laudable pus. On this, the pain and swelling quickly ceased, and, on the 28th of October, a loose piece of the humerus, of the size of a nut, and plainly a part of the rotula, was removed. The orifice of entrance cicatrized in the beginning of November, the wound by operation was filled with exuberant granulations, necessitating applications of a solution of nitrate of silver. Through the last made incision pus still flowed abundantly. On the 18th of November, an incision was again requisite on the volar side of the fore-arm, on account of gravitating abscess. Applications of warm poultices quickly reduced the suppuration, and by the beginning of December, the wounds were almost healed. The patient now left his bed, and careful movements of the arm were daily attempted. On the 17th January, 1851, I saw him again in the Convalescent Hospital in Jevenstedt. The orifice of entrance of the wound had again broken out, and it appeared as if another splinter was coming away. The remaining wounds were cicatrized, but the elbow-joint was entirely anchylosed.

Case II.—Splintering of the Olecranon. Resection of this Part of the

Bone after 14 days. Cholera. Abscesses. Tedious Healing, with perfect Anchylosis—In the battle of Idstedt, on the 25th of July, 1850, the Schleswig-Holstein private S——, received a rebounding-shot from an apparently spent ball, on the right olecranon, and came, the same day, into hospital at Altona. After the first inspection it was deemed a superficial grazing of the olecranon, the wound exhibiting no symptoms of inflammation of the joint.

Towards the end of the second week we first remarked that, on pressure on the front of the elbow-joint, a synovial-like fluid proceeded from the wound; on closer inspection, and after a slight dilatation of the wound, a star-like fracture of the olecranon was discovered, it being possible to introduce a small probe between the separate fragments. Considering that the joint had been already opened, and that if the cure was to be left to nature, protracted exfoliation was to be expected, Dr. Ross proceeded, 14 days after the injury, to the removal of the olecranon. It was laid bare by an incision across the middle, the soft parts separated laterally; it was then freed from the biceps and the ligamentous structures, and then, with a slight flexion of the fore-arm, was sawn off at its root with a small, strong convex saw. The separate pieces of the freely-splintered bone only hung together now by ligamentous bands. There was no further injury, the exposed cartilage of the humerus was unaffected. The wound was covered with oiled charpie, and the arm slightly bent and laid on a splint. The bandage was not changed for some days, and the patient had no pains in the joints or bad symptoms. The wound was already covered with such florid granulations that they hindered the view into the joint, when the patient had an attack of Asiatic cholera, at that time prevalent in the hospital. He did not succumb, but tedious diarrhœa came on, and he became very thin. While in this condition an abscess formed deeply in the front of the elbow-joint, accompanied by much pain and fever. It was opened rather tardily. The patient recovered by degrees; the elbow-joint regained its natural condition; the wound by operation was closed, but not cicatrized. Some flexion and extension of the fore-arm was now tried, which again brought on pain and fever, and an abscess again formed in front of the joint. When the patient slowly recovered, and the wound was entirely cicatrized, no further movements of the elbow-joint were attempted, but only those of the hand and fingers.

When he left the hospital, in February, the fore-arm was fixed by false anchylosis at an angle of about 135 degrees with the upper arm, in slight pronation. The movements of the hand and fingers were perfect and powerful.

Another question is to be considered, viz.; how much of the comminuted bone may be taken away, and in what extent of comminution it can be fairly attempted to preserve the limb by resection. We have proceeded much farther in our cases, than Guthrie and Baudens recommend,—they would confine this operation to some certain cases, yet, in truth, neither speak from experience. Guthrie would only resect, where the lower end of the humerus alone, or where the upper joint-ends of the ulna and radius are together injured. Baudens is of opinion that resection should only be performed where one alone of the three joint-ends is injured. We have performed this operation, where not only one or two, but all three bones of the elbow-joint have been injured. We have removed, from their surrounding soft parts, four to five inches in length both from the humerus, ulna and radius, when these portions had been struck off by the bullet. Indeed, in one case, where a resection of the shaft of the humerus in its lower third had been commenced, the elbow not being supposed to be injured, such a considerable extent of injury was found, that the epiphysis of the humerus was also removed;—this case proceeded as favorably as the others, and, after the arm had become much shortened, it healed with almost complete anchylosis, although, in all, a piece of four inches in length of the entire volume of the humerus had been removed. If the joint-ends had been struck off obliquely, we never sawed off the whole portion of the shaft thus injured, but, on the principle that a broken surface of bone would not be more dangerous in a wound than a sawn surface, we merely truncated the projecting portions. For instance, when we were obliged to remove very long pieces of bone, obliquely struck off the shaft,—we removed the sharp ends remaining in the wound, by a small saw; and we have frequently been able to observe, that no large sequestra were afterwards discharged from the broken surfaces, but, on the contrary, that, soon after cicatrization of the wound, considerable masses of callus were thrown out from the bone, which strengthened it and was of great use in rendering the arm firm.

## Method of Operating.

As regards this point, we have tried very many different methods; but in the last campaign we operated solely in Liston's manner, with certain modifications, which partly, rendered it essentially more easy of performance, partly, was necessitated by the varying extent of the injury to bone. If the splintering of the bone is slight, and resection is only demanded by the suppuration of the joint, the operations can be performed as on the dead body; if the joint-ends, however, are much comminuted or entirely separated from the shaft the operation would be modified on these accounts.

All methods in which osteotomes, chain-saws, and other complicated instruments are necessary, are of course not to be used in war, partly, as these instruments are so costly, so that many surgeons are unable to purchase them, partly, because they very quickly get out of order, and can only be repaired by an instrument maker. To resect the elbow-joint quickly, and with the simplest instruments, it is first requisite to lay open the whole joint so freely that the ends of the bones can be easily thrust out of the wound. This is only possible by opening the joint on the posterior surface, as there is here no other important vessel or nerve, except the ulnar nerve. Earlier operators have at once divided this nerve; and doubtless, in many cases, reunion with restoration of function has taken place. As this can, however, be easily avoided, it is absolutely requisite, for performing this operation in a scientific manner, to preserve this nerve, unless indeed the internal condyle of the humerus is comminuted, and the nerve torn across at the same time.

For this operation, the patient was laid on his back, on a table covered by a mattress, with the posterior surface of the wounded arm turned to the light, and lying on a long cushion covered with oil-cloth. If the patient lays prone, the operation is certainly easier, but we preferred the method described on account of its being less dangerous to administer chloroform. An able assistant held the arm, a second is required to move the fore-arm according to the wishes of the operator. As soon as the patient was under the influence of chloroform, an incision, three inches in length, was made on the outer side of the ulnar nerve, to divide the skin and subcutaneous cellular tissue. This incision must begin just above the point of the olecranon on the inner side of the tendon of the triceps, proceed along the inner edge of the olecranon—and below the same, corresponding to the direction of the crista ulnæ, be carried a little more outwardly. The second incision is carried from over the humero-radial joint transversely over the olecranon into the middle of the first—meeting it at a right angle. Earlier operators have now opened the sheath of the ulnar, and drawn it to the inner side by means of a blunt hook. The operation is thus much easier and even a less practised surgeon does not run the danger of wounding the nerve. The Surgeon in Chief, Dr. Stromeyer, in general therefore recommended this method to younger surgeons performing the operation for the first time. The nerve is thus, however, injured. If the operation is performed with suitable care, this nerve is not at all seen—by adopting the following procedure.

After the incisions in the skin, the fore-arm is somewhat flexed, and the capsule of the joint opened at once on the inner side of the olecranon. The edge of the capsule is now either seized by hooked forceps, or the edge of the wound is drawn by the left thumb sufficiently inwards, for its nail to stretch the incised portion of the capsule. Repeated incisions are now carried along the inner border of the olecranon direct upon the internal condyle, by which the soft parts are entirely separated from the bone. These incisions must follow one another at the most at the distance of half a line, and the edge of the knife must always cut on the bone; for by this means only can the periosteum covering the inner condyle be left in connection with the capsule, and all the soft parts above it be dissected back without injury.

The knife is generally blunted during this stage of the operation, but when this is ended, the greatest difficulty is overcome. This method is especially necessary in all resections, if the periosteum is to be preserved and the soft parts injured as little as possible. As the exposure of the inner condyle is the point to be effected by any one wishing to become master of the operation, its practise on the dead body is especially to be recommended to every surgeon.

When the internal condyle, which often projects very far, is laid bare to its point, the soft parts are to be drawn away by the left thumb, and the periosteum from it as before described for the olecranon. After this, the position of the ulnar nerve may be recognised by a prolonged swelling on the inner side of those soft parts turned back from the condyle;—but the sheath of the nerve itself cannot be seen, as it is covered by a portion of capsule, of periosteum, and by a thin layer of cellular tissue.

The third stage of the operation is now proceeded with,—that is, the complete opening of the joint. The internal lateral ligament, already laid bare, is first divided in its middle. The arm being then much flexed the tendon of the triceps is freed by the knife from the olecranon, and the incision prolonged outwards and downwards to the head of the radius, so as to divide the anconeus from the ulna and also to open the capsule in this position. By means of a transverse incision, the external lateral—and the radial annular—ligaments are divided, and the head of the radius exposed. The whole joint now gapes widely on strong flexion of the fore-arm so as to touch the arm—and then by taking hold of the olecranon and drawing it from the humerus, one is generally at once able to divide the anterior portion of the capsule with the knife. If the joint-ends do not yet separate sufficiently, it depends upon the lateral ligaments not having been fully severed; the portion in fault, whether internal or external, must be sought for by the fore-finger, and at once divided by the knife. If a large portion of the ulna is to be removed, the attachment of the brachialis anticus must be separated from the coronoid process, taking care not to wound the brachial artery; we have often used Langenbeck's blunt-pointed knife for this portion of the operation,—if care is taken to cut only against the bone, a pointed knife is equally serviceable for the purpose.

The joint-ends of the three bones can now be thrust far enough out of the wound so as to be removed by an ordinary amputating saw. In cases, where there was but a slight injury to bone, as for example, when only the olecranon was comminuted, or the joint-surface of the humerus grazed, we sawed off the head of the radius, the epiphysis of the ulna close below the coronoid process, and the whole end of the humerus covered by cartilage. If it was necessary to take off a larger piece from one bone, so much less was removed from the others. If, for example, the lower end of the humerus was comminuted to such an extent, that, after removal of the splinters and truncation of the sharp points, three to four inches were found to be lost,—we left the ulna and radius untouched; if the loss of substance was less, the olecranon, or only the upper half of the same was sawn off, as its projection, against the end of the resected humerus, easily induces pain and increases the probability of anchylosis. If we were obliged to resect large portions of the ulna and radius, we left the end of the humerus similarly untouched.

At first we occasionally attempted to remove the cartilage with a knife; this is troublesome, however, and wastes time and cannot be effected completely, moreover, as it appeared to have no influence on the further course of the case, we did not do so latterly. The cartilage softened soon after the occurrence of suppuration, and freeing itself in larger or smaller pieces, came away with the pus from the wound.

I have already mentioned that in the year 1850, we only resected the elbow-joint after Liston's method. In the campaign of 1849, Dr. Stromeyer at first resected after Jaeger's manner, as being the easiest to show to the younger surgeons; this consists in H-formed incisions, the two flaps of which are dissected back. After the sheath of the ulnar nerve was opened, this was drawn to the side and the operation then proceeded with, as already described. In three cases we added another incision to Liston's ⊣-formed one, by making a short one parallel to the longitudinal one, commencing above the humero-radial joint and carried either upwards or downwards. This was done according as the humerus or ulna were comminuted. The wound thus became ⊣-shaped, and the removal of a large piece of bone was much facilitated. By this means Liston's method of incising the skin sufficed for all cases, and the method already described was later always employed, viz.: in twenty-seven cases.

The Surgeon-in-Chief, Dr. Langenbeck, preferred to make a single longitudinal incision, three inches in length, on the inner side of the olecranon, corresponding to the course of the ulnar, and the method of dissecting back this nerve from the internal condyle was first taught by him. If the soft parts are not yet much swollen, the ends of the bones may be easily thrust out from such an incision. As soon, however, as a considerable serous or inflammatory infiltration has come on, as generally occurs after gun-shot injuries, it is necessary to lengthen the incision too much, to prevent tearing or bruising the skin. This method was employed three times.

In a case where the lower end of the humerus was comminuted, Dr. Stromeyer made a semicircular incision, the convexity downwards, in the same way as is given by Guepratte. This method has no especial advantage and the exposure of the internal condyle becomes somewhat more difficult.

If the ulna is struck below the joint and the splintering extends into the same, and to such an extent that it is unnecessary to remove any portion of the humerus,—it is not requisite to expose the internal condyle. In such cases the joint may be opened outside the olecranon, and the fragments be dissected out one after the other. The internal lateral ligament may be divided from the joint-cavity—merely taking care to avoid the ulnar nerve. Similarly all the soft parts over the epiphysis of the humerus are as little as possible to be meddled with.

If the resection is undertaken some time after the reception of the wound, the hæmorrhage is at first free, as the smaller arteries of the skin and cellular tissue are much enlarged by the inflammation. However, we never compressed the brachial artery, as the bleeding could only be of service to the patient, and, besides, would generally cease during the operation.

After completion of the operation, the centre of the longitudinal incision was usually left open for the escape of matter, and the remainder of the wound united by interrupted sutures. We have not seldom seen a partial union of the wound follow by first intention.

If, however, one or both gun-shot wounds were on the posterior surface of the elbow, we at once changed the direction of the incisions so that they should run through the openings, and then left this portion of the wound open, as we could not expect that the contused borders would heal by the first intention.

## After-Treatment.

Stromeyer was certainly right in considering the quiet and comfortable position of the arm as one of the most important points of the whole after-treatment. By his direction the arm was laid at once, after the operation, on a smooth, wooden splint, padded with wool and covered by oil-cloth. This reached from the upper third of the arm to the finger-ends, and, at the elbow, was in an obtuse angle of 140 degrees. A hole, the size of a crown, was cut in the splint to allow the internal condyle of the humerus, if left, to be protected from injury. This splint lay near the patient on a large chaff cushion, the arm was laid prone upon it, and kept in position by some turns of a roller above and below. The wound was dressed at first with cold, afterwards with warm water, when the suppuration diminished, charpie dipped in ordinary oil was applied.

Stromeyer laid great stress upon the arm never being raised from the splint during this treatment. The charpie was laid around the wound sufficiently to take up the secretions from it, and, if care is taken for this object, the arm can be kept thoroughly clean. It is only necessary to dry the under surface of the arm from pus, by small masses of charpie or sponge, by means of the spatula and forceps.

If purulent deposits form deeply, their contents must not be emptied by pressure towards the wound, as the evil is thus increased; but poultices and early incisions must be employed in suitable spots, so as to allow free issue of the matter. If the arm becomes œdematous, bandages are of great service, of course, however, many short pieces, of a foot to a foot and a-half long, must be employed, so that the arm shall never be raised from the splint. It is only when the wound is filled with granulations, and cicatrization has already commenced,—that the arm may be carefully raised from the splint, and a roller be employed, a flannel one is best for this state of the arm. At this time, also, warm baths for the whole arm are of great service; by this means, especially, the freedom of motion of the arm and fingers is greatly aided.

The effect of movement in the production of false joints after fractures is well known; it is in the same way of great importance to institute passive motion of the elbow sufficiently early, in order to ensure the usefulness of the arm. It may be commenced with care, before the wound is fully cicatrized, but, as soon as irritation is caused and the granulations acquire a worse appearance—it must be omitted or very severe inflammatory symptoms or hæmorrhage will easily occur.

If all the wounds are fully cicatrized, the patient must make up his mind to bend, extend and rotate the arm frequently by means of the sound one, and the surgeon himself must perform the same passive motions of the arm whenever he makes his visit, or cause it to be done by an assistant or by another patient. As this practice is painful, the patient readily omits it, even when directed to do so. I believe it is to this circumstance chiefly that must be ascribed, the more or less anchylosed joints that frequently occurred. As, in war-time, the surgeon who first attended the case, easily looses sight of it, often before the wounds are fully cicatrized, by the patient being transferred to a distant hospital, where the surgeon is not so attentive in ensuring motion of the joint, as the operator himself, who wishes that as good a result as possible may be obtained from his operation.

The great importance of proper after-treatment, as regards the later usefulness of the limb, was decidedly shown in patients resected in the elbow by us at Schleswig, after the battle of Idstedt, and who later were treated in the Danish hospital. The Danish surgeons have not, to my knowledge, performed this operation, and the majority of them, therefore, were not likely to know the value of passive motion in such cases. Hence, it happened, that when these patients returned from captivity, in the beginning of 1851, most of them had their arms in a very bad condition, and in part possessing neither motion nor proper sensibility. These evils could be partially bettered by means of warm baths and methodical passive motion. Yet the mobility of the arms of those patients resected after the storming of Friedrichstadt—that is, almost two and a-half months later than the others—had advanced much farther at the time of the return of the former.

As the mobility of the fore-arm after the operation, does not wholly depend upon the size of the portions of bone removed, it will be understood, that the joints, in some patients from whom portions of considerable extent had been taken away, nevertheless resulted in anchylosis, while great freedom of motion existed in others from whom but small portions had been removed. On the whole, it is preferable to take away more rather than less, as the probability of later mobility is thereby increased. In general, therefore, we removed the cartilaginous extremities of all the bones, even if but one of them had been slightly injured.

The greater number of patients with anchylosis of the joints, did not have the arm at a right angle, but at an obtuse angle of 130 to 140 degrees. This is because this position is the only comfortable one for the patient while in bed. We have frequently observed that, when the fore-arm was more flexed, the patients complained of excessive pain, which only ceased on the arm being more extended.

## Results.

Of 40 patients for whom resection of the elbow-joint was performed in the three Schleswig-Holstein campaigns; 6 died, 1 is not yet healed; in 1 the fore-arm mortified, and it was necessary to remove it by amputation, the remaining 32 are completely healed and have a more or less useful arm. As regards 2 of them, I have not been able to learn anything with reference to the power of motion they possess, of the rest 8 have very extensive—9 more or less complete—power of motion; it is to be hoped of many of the remainder, that they will be able to obtain much increased mobility by means of zealous exercise of the arm. On the other hand, 13 of the cases have a more or less complete anchylosis of the joint.

As regards the space of time between the reception of the injury and the operation, a similar relation exists, as in resection of the shoulder-joint, and indeed as in amputation of the larger limbs. Of 11 cases in which resection was performed in the first twenty-four hours, only 1 proved fatal. Of 20 cases, in which it was necessary to perform the operation in the condition of highest inflammation and commencing suppuration; that is, from the second to the fourth day, 4 cases were fatal. Of 9 secondary resections, which were performed from the eighth to the thirty-seventh day only 1 ended fatally. Here, also, therefore, it is an important rule, to perform the operation, as soon as possible, after the injury; if the inflammation, however, is already very considerable, and the suppuration but commencing,—it is far preferable, first, to moderate the inflammation by antiphlogistics, and only to operate when the suppuration is free and the infiltration of the soft parts has somewhat diminished, if indeed it may be ventured to protract the operation so long.

We were unfortunately obliged in the majority of cases to put off the resections over the twenty-four hours, for, on the first day after a battle we were generally compelled to give our whole time for the amputations in which every delay was of far greater danger. That strict antiphlogistic treatment is of the greatest use in lessening the danger, is shown by the results of the two last campaigns. In the year 1849, we could rarely obtain ice, and so we find that of the 4 fatal cases, operated on from the second to the fourth day, 3 of them occurred in this year. In the year 1850, we generally had sufficient ice, so as to moderate the inflammation for this operation, and hence but 1 fatal case occurred of all the resections performed; even this case, I do not think, should be directly attributed to the operation, as the patient went on perfectly well, as long as he was treated by our surgeons, and the wound was nearly healed, when he came under the care of Danish medical men. They later performed amputation of the arm, and the patient did not survive the operation. I have not been able to learn, what cause rendered amputation necessary.

It is to be hoped, that after these results of resection of the elbow-joint after gun-shot wounds, the practice may become more and more common among army-surgeons, and that in future the arms may be preserved to many warriors, in whom formerly amputation was at once performed.

I give here briefly the history of those cases in which resection of the elbow-joint was performed. [Of these cases nine are translated.]

Case II.—Comminution of the internal Condyle of the Humerus. Resection on the 19th day. Healing.—The Danish Major von W.——, aged 56, of slight stature and suffering from chronic cough and asthma,—was shot in the left elbow at Oeversee, the 24th of April, 1848. The joint was opened on its outer side, and the external condyle comminuted. On the 10th of May, phlegmonous inflammation had spread to the shoulder, so that the surgeon in attendance was thinking of exarticulation of that joint. The Surgeon-in-Chief, Dr. Langenbeck, proposed resection, and performed that operation on the 13th of May. He made one longitudinal incision, dissected back the ulnar nerve and removed, by the saw, the ends of all the three bones. The great swelling of the soft parts added much to the difficulty. After the patient had recovered from the effects of the chloroform, it was evident, that both sensibility, in the course of the ulnar, and power of motion of the hand and fingers was still possessed by him. Complete cicatrization took place about the end of August, and the arm already was capable of some extent of mobility,—the patient now returned home.

Case III.—Communition of the Joint-end of the Ulna. Resection the next day. Healing.—The Hanoverian private, K——, was shot through the right elbow by a musket-ball, on the 6th of April, 1849, in the skirmish at Ulderup. The upper end of the ulna was much comminuted. The Surgeon-in-Chief, Stromeyer, resected the joint on the following day. He employed the H-formed incisions through the skin, on purpose to shew to the numerous young army-surgeons present, the easiest method of performing the operation. One longitudinal incision was over the outer—the other over the inner—condyle, the transverse one over the upper part of the olecranon. The ulnar nerve was now dissected back, and drawn aside by a blunt hook. After division of the triceps tendon, the joint was opened, the free splinters of the ulna removed, and the sharp points of the same, and the head of the radius sawn off. In a similar manner, the greater portion of the cartilaginous surface of the humerus was taken away by the saw. The skin was united by some interrupted sutures. Recovery proceeded slowly, but without any particularly bad symptoms. I have not been able to learn whether complete anchylosis followed, or whether mobility remained in the elbow; at any rate the patient was able to use his hand well, when he left the hospital.

Case V.—Comminution of the Lower Joint-end of the Humerus. Resection on the 2nd day. Amputation of the Arm. Death by Pyæmia.—The Saxon rifleman, R——, was wounded on the 13th of April, 1849, in the left elbow by a bullet, at the entrenchments at Dueppel. He was not brought to a hospital at Flensburg, until the following evening. The joint was already luxated and fluctuating. On the 15th of April, resection was performed by Dr. Langenbeck, who was then making his visit at Flensburg. A longitudinal incision, of four inches in length was made, the ulnar nerve turned aside, and then the joint opened. A large quantity of thin synovia and coagulated blood escaped. The whole epiphysis of the humerus and the olecranon were removed by the saw, and the skin brought together by a number of interrupted sutures,—allowing, at the same time, an opening in the centre for the escape of pus. No splint was employed. At first the case went on favorably,—on the fifth day, on removal of the sutures, the greater portion of the wound was already healed by first intention,—the pus was laudable and not too profuse. Towards the end of the month, however, the granulations became flabby, and on the inner side of the arm, in the neighbourhood of the vessels, there appeared a hard swelling which soon gave decided fluctuation. On dressing the wound, on the 30th of April, some blood was found mixed with the matter. On the 1st of May, slight hæmorrhage occurred on dressing the wound in the morning, but it soon ceased spontaneously; in the afternoon, however, it recurred, and to such an extent, that the patient had soon lost two pounds of blood. Ligature and torsion were vainly tried,—probably the blood flowed from numerous small vessels, and the hæmorrhage was the so-called pyæmic, due to the larger venous trunks being blocked up. Amputation was, therefore, resolved on, and performed by my father who had direction of the hospital. On examination of the arm, both resected ends were found covered with callus. The source of the hæmorrhage could not be discovered. After the amputation the patient became rapidly worse,—rigors occurred on the 4th of May, recurred twice daily and were followed by profuse sweats. However, the aspect and suppuration of the stump remained apparently good. On the 9th of May, the patient first complained of shooting pains in the chest, dyspnœa and delirium followed. and, on the 14th of May, the case proved fatal.

On the autopsy, numerous pyæmic abscesses were found in the lungs, liver and spleen, with recent exudation in the left pleura.

Case VI.—Comminution of the upper Joint-end of the Ulna. Resection on the 2nd day. Healing.--The Schleswig Holstein private, Christian B——, had his right ulna comminuted by a bullet, one inch below the point of the olecranon, in the battle of Kolding, on the 23rd of April, 1849. He was brought to Hadersleben on the following day—a loose splinter was extracted from the entrance-wound. Dr. Stromeyer saw him on the evening of the 25th, and, on examination, recognized injury to the joint. He, therefore, at once proceeded to resection. The operation was performed by artificial light, with the assistance of a Professor of Antiquities and a legal officer, besides that of myself and the surgeon of the hospital. The H-formed incisions were employed, the splinters of the ulna removed, and the sharp points of the lower fragment and the head of the radius taken off by the saw. A portion of the cartilaginous investment of the humerus was removed by the knife. After insertion of the sutures, the arm was laid on a padded splint, on which it remained without being moved, until the wound was fully healed. This took place about the end of June, without intervening accidents,—passive motion was now employed. In July the patient mixed in a fight while drunk in an ale-house, using both his arms vigorously. The wound broke out afresh and required some weeks before again cicatrizing. Passive motion was again actively employed. When Br——, left the hospital, on the 25th of August his arm was in the following state :—the right fore-arm was slightly thinner and weaker than the left, the right hand was perfectly strong and free in its motion. The upper end of the radius was somewhat displaced forwards and inwards, but pretty well bound to the ulna, so that it got but little out of position with the motion of the arm,—pronation and suppination were almost perfect. The arm could be flexed by the patient to a right angle, extension was only possible by means of the weight of the fore-arm.

Case VII.—Splintering of the Internal Condyle of the Humerus. Resection on the 3rd day. Gangrene of the Fore-arm. Pleurisy. Empyema. Recovery. — The Schleswig-Holstein, Acting-Officer Max v. F——, was struck by three bullets, at Kolding, on the 23rd of April, 1849. One pierced the soft parts of the left thigh, another the superficial layer of the abdominal muscles, the third was driven into the internal condyle of the right humerus, and was not to be found. On the following day there was extensive swelling about the elbow, which, in the few next days, rapidly passed up the arm, and was so marked that blebs formed here and there up to the axilla. On the 26th of April, Dr. Stromeyer visited Christiansfeld and selected this one, as well as the two following, as fit cases for resection. I performed this after Liston's method. On incision of the skin, more hæmorrhage occurred from the cutaneous vessels than we had seen before in this operation. After separating the soft parts from the internal condyle without seeing the ulnar nerve—which, indeed, had been divided by the bullet—the joint was opened, and the epiphysis of the humerus and the olecranon was sawn off. Directly after the operation we noticed that the fingers of this hand were strongly flexed and could not be extended voluntarily—the hand had lost sensation, and no pulsation could be discovered in the radial or ulnar arteries. This could not have been due to the operation, as it was performed with the greatest care. Unfortunately a careful examination of the sensibility and of the pulse, had not been previously made, on account of the number of severely wounded to be attended to. The bullet was not found during the operation ; probably it had torn across or crushed the vessels and nerves at the bend of the elbow, so that the collateral vessels being divided by the requisite incisions, all circulation in the arm was interrupted. We took more care after this case, to examine all the conditions present in the closest manner before an operation of this kind.

On the following day, the first symptoms of mortification of the arm occurred with the phenomena of septic fever. The patient was so weak, that amputation could not be thought of. Gangrene spread quickly to the elbow, the line of demarcation formed here, and, on the 3rd of May, the limb, which had already become quite black, could be removed near to this line, without any hæmorrhage from the stump. From this time the patient advanced more and more favorably, the remaining mortified portions soon fell off, the irregular flaps soon began to form granulations, and the other wounds went on well. On the 11th of May, violent pleurisy occurred on the right side. The effusion of empyema broke out under the scapula, and by incision, on the 30th of May, a large quantity of pus was discharged. The patient, who had, for the second time, become extremely weak, now again gathered strength, though very slowly. The healing of the stump formed by nature, was very tardy, as the large quantity of soft parts, at first present, had now retracted so much, that a portion of the humerus was exposed. Nevertheless, although the end of the bone is merely covered by cicatrix, the patient makes no complaint of it, and is again in service, as an officer. It was not till the following spring that F——, had recovered his full strength.

Case XI.—Comminution of the Lower Third of the Humerus. Resection of a piece four Inches in length. Healing.—The Danish private Thomas S——. had his left humerus comminuted by a bullet, two and a half inches above the elbow, at Kolding, on the 23rd of April, 1849. In consequence, the arm swelled very considerably, and, on the 25th inst., Dr. Niese, after dilatation of the entrance-wound, removed the loose splinters and truncated with a saw the sharp points of the upper and lower fragments. In this manner some two inches in length of bone was taken away. At first, the elbow was thought to be uninjured, but, even the day following, the increase of inflammation in the same, showed that a fissure must have extended to the joint. The lower fragment of the humerus, two inches in length, was still remaining, and this was removed on the 27th of April, by prolonging the first longitudinal incision. The case proceeded favorably; at the end of July the patient was recovered and could be discharged from the hospital and exchanged. In spite of so great a loss of substance, there was so little passive motion, that it was almost complete anchylosis. The hand was strong and useful.

Case XIII.—Comminution of the Head of the Radius. Resection on the first day. Healing. Ensign Heinrich S——, of the 6th Schleswig-Holstein infantry battalion, was shot at Fridericia, on the 6th of July, 1849, in the left elbow. On examination, the head of the radius was found comminuted, I, therefore, resected the joint, after Liston's method, on the following day. The splinters of the radius were removed and the ulna sawn off at the same height, only a portion of the cartilage was taken off the humerus. The case was treated as already described; every thing went on well to the third week, when the granulations bled to some extent, probably from tossing in sleep and moving the limb. One of the acting-surgeons was preparing for amputation, when, fortunately, Dr. Stromeyer, who was on his visit, interposed. Hæmorrhage did not recur, and recovery took place speedily, so that the patient could be discharged, at the end of three months from the first. The joint was anchylosed, yet it could be used so well, that S——, again took service in spring 1850, and went through the greater portion of that campaign as ensign, until, in autumn, he received a civil post.

Case XIV.—Comminution of the Head of the Radius and of the Upper End of the Ulna. Resection on the first day. Death by Pyæmia. The Schleswig-Holstein private P. Q——, was shot in the right elbow at Fridericia, the 6th of July, 1849, and brought to Christiansfeld. The bullet had penetrated the fore-arm transversely, and had comminuted the head of the radius and the ulna close below the coronoid process. The arm became very much swollen during the transport, and the operation could not be performed until the following day, as he was brought to the hospital in the evening, together with a large number of severely wounded soldiers. On the 7th of July, therefore, Dr. Harald Schwartz performed resection after Liston's method with removal of the joint-ends of the ulna and radius. The splintering of the ulna extended far downwards,—the piece splintered off was three and a half inches long. As the fracture, however, was oblique, the lower fragment reached up to the neck of the radius and only its sharp point was removed by bone-forceps. The swelling of the arm at first decreased, but soon recurred and extended by the side of the vessels up to the axilla. Blebs of clear serum formed on the reddened and tightly distended skin. The violent fever soon was accompanied by delirium and a dry tongue, and death occurred, on the 10th of July—the third day after the operation. On examination the lymphatics and small veins of the arm were found full of pus, which oozed from many small openings, on incising the soft parts. The lungs had numerous pyæmic abscesses most of them still in the state of hepatisation. The other organs were sound.

Case XXXV.—Comminution of the Lower Joint-end of the Humerus. Resection on the 10th day. Healing. — The journeyman tanner, Carl S——, on the 7th of August, 1850, trying to save his master's child, who was playing in the street, when the arsenal exploded at Rendsburg, was struck on the right elbow, head and back either by falling fragments or pieces of a bomb. He was first treated by a civil surgeon, and was brought into a hospital on the 14th of August. The wounds of the head and back were then nearly healed, however, the elbow was extremely swollen and tender, and a large quantity of sanious matter came out from a lacerated wound, which ran transversely close above the olecranon. On close examination, the end of the humerus was found divided in two pieces, and separated from the epiphysis. Dr. Kunkel, Assistant-Surgeon, therefore, performed resection on the 17th of August,—he removed the loosened fragments of the humerus and sawed off the jagged broken ends; in all a piece of two and a half inches long, was removed. The patient had, immediately after the operation, a numbed sensation in the course of the ulnar nerve; it had been seen that this nerve in its sheath had been displaced, but not divided, by the force causing the injury. At first, after the operation, the patient had violent pain in the wound, this soon ceased, and the general state becoming more satisfactory day by day, the wound was nearly healed on the 18th of September. At a later period, matter formed at the inner border of the biceps, it was opened, but the complete recovery was so far hindered, that the man was not discharged until December. On January the 17th, 1851, I saw the patient again. He was using baths, and the mobility of the arm was increasing, he could already flex and supinate the fore-arm slightly, the hand was in use, but the fingers still somewhat stiff. Sensibility was recovered in the course of the ulnar nerve, yet, as soon as the arm became cold, numbness and formication was felt. According to the description of the patient, the nerve must have been displaced anteriorly.

### c. *On Injuries of the Hip-joint.*

Injuries of the hip-joint by bullets are not very frequent, on account of the deep and protected position of this joint; yet they occur more often in war than Larry describes, "Chirurgische Klinik." He states to have not seen a single case of the kind, although of all authors, he has given the most extensive observations on gun-shot wounds. This is, probably, explicable from the majority of these wounds proving rapidly fatal, and from the time being so occupied after a battle, that surgeons have no time to perform post-mortem examinations.

We have frequently had the opportunity, in the Schleswig-Holstein campaigns to observe wounds of the hip-joint, generally, however, they were complicated with other severe injuries.

Comminution of the os femoris alone is scarcely possible, on account of its protected situation in the acetabulum, if so, the bullet must have penetrated the surgical neck, from beneath, and lodged in the head. An injury of the capsule alone, may occur, if either a spent ball has passed in sufficiently far, or, if one, traversing the limb with full force, grazes the capsule or fibro-cartilaginous edge of the acetabulum; in these cases contusion of the bone is generally present.

We had a fatal case of the last description after the storming of Friderichstadt; the bullet had injured the tuber of the os ischii, as well as the capsule. Pyæmia rapidly followed the profuse sanious discharge.

The acetabulum may be injured from the outside as well as from inwards. After the battle of Idstedt, a case occurred to us, where a conical bullet had penetrated through the great sciatic notch into the pelvis, and had pierced the posterior boundary of the acetabulum. Here it lodged and projected, with its lateral convexity, into the joint-cavity. The os femoris was scarcely contused, but, of course, sanious discharge occurred from the joint, together with a large collection of blood and matter in the pelvis, this rapidly caused a fatal termination.

All these injuries are very difficult to diagnose accurately, on account of the position of the part, being generally too deep to reach by the finger, and, as generally there is no characteristic alterations in the limb. One can only conclude what kind of injury exists, by the extensive swelling occurring rapidly, and by the great pain felt on motion. As in most cases, the pelvis is likewise injured, as the injuries of the spongy tissue of these bones, if severe, are almost always fatal by pyæmia, it is easily understood, that the surgeon, in these cases, can scarcely think of an operation, and must content himself with allaying the sufferings of the patient by suitable position, cleanliness, antiphlogistics, opium, &c.

It is otherwise with injuries of the neck of the femur, or of the neighbourhood of the trochanters; these are not unfrequent, but often very difficult of diagnosis. When the bone, in this neighbourhood, is comminuted, the fragments are sometimes bound together, at first, so thoroughly, that the usual signs of a fracture are not present. The limb is not shortened, nor the point of the foot turned outwards and, for the first few days, the patient himself may move the limb without very much pain. We have seen many cases, where the extent of the injury was only known on the occurrence of suppuration, and after symptoms of pyæmia had already shown themselves.

If the case is left to nature, it is generally fatal, after the patient has suffered much and severe pain. For, either fissures extend from the comminuted spot of the neck into the joint, and inflammation of the joint occurs, after the suppurative process has passed along the fissures from the wound; or, if the neighbourhood of the trochanters is comminuted, the large proportion of spongy tissue in the bone, offers so favorable a focus for the generation of sanious pus, that the whole medullary substance is soon involved, and pus is taken up into the blood by the veins of the bones. Such cases seem to demand an operation in the first instance,—the question lies between exarticulation or resection of the hip-joint.

Exarticulation was performed, in all, 7 times, in the Schleswig-Holstein campaigns, 5 of these were operated on by B. Langenbeck. Only 1 of these patients, a young man, seventeen years of age, recovered, This result could not encourage us to undertake the operation; again, the indications for its performance frequently manifested themselves with certainty, when it was already too late.

Stromeyer, therefore, allowed resection of this joint to be performed in 1 case, and, although this man died, yet, in similar cases, this operation is, I think, to be preferred to exarticulation.

The Danish private O——, was shot in the left hip, at Kolding, on the 23rd of April, 1849, the great trochanter was comminuted, and the bone fractured obliquely, through the two trochanters. Local and general symptoms were at first so slight, that the attempt was made to leave the case to nature, by proper position and antiphlogistic treatment. In some weeks, however, the discharge became sanious and profuse, but, as it could not escap freely, it was determined to enlarge the wound and to remove all the splinters which might be fully loose. On the 13th of May, Dr. Harald Schwartz dilated the entrance-opening—directly upon the great trochanter—longitudinally upwards and downwards for four inches, and then extracted many small fragments by dressing forceps. On further examination, it was found that a fissure extended along the neck of the femur into the joint, the upper fragment, therefore, was seized by Langenbeck's bone-forceps, the capsule and ligamentum teres divided and the head exarticulated. The lower fragment was now thrust out of the wound, and the injured portion, to two inches below the small trochanter, was sawn off. Little difficulty was experienced. At first, after the operation, the patient improved in condition, the fever decreased, and we were already hoping a favorable termination, when, suddenly, on the third day, a violent rigor occurred, repeated daily; metastatic inflammation came on of the right foot and shoulder-joints, and death followed on the 20th of May. On the autopsy, it was found that the tuber ischii was also injured by the bullet, and that its spongy substance was infiltrated with sanious fluid. The right foot and shoulder-joints were full of pus, the internal organs were sound.

Dr. Ross had a similar operation, but performed two years after the injury, the case is fully described in No. 41 of the "Deutsches Klinik, 1850." It also ended fatally; if performed sooner, it is possible that life might have been saved.

It cannot be denied, that the wound of resection is far less in extent than that of exarticulation. It must be decided, in future campaigns, which operation should be performed, and when. I should prefer resection in all cases in which the vessels and nerves were uninjured, and where there was no considerable injury of the pelvis.

Comminution of the great trochanter, without fracture of the femur, occurs occasionally, and the spongy structure of this portion is the source of great danger to the patient. The strictest antiphlogistic treatment must be employed, and no incision nor dilatation be made until required by suppuration. All attempts to extract the splinters early are very dangerous, according to our experience, as it gives the air more free entrance to the deep and superficial wounds, and aids the extension of inflammation to the joint.

### d. *On Injuries of the Knee-joint.*

All that we have said of the unfortunate results of injuries of the joints by gun-shot, applies especially and in the highest degree to the injuries of the knee-joint. All authors concur in its extreme danger, and we have had the sad experience of the same, as they are very common in battle.

Injuries of the capsule alone, are, on the whole, unfrequent, yet they are proportionably more common in the knee than in other joints, on account of the greater extent of the capsule beyond the bones. As violent inflammation and suppuration are imminent, amputation of the thigh is generally necessary sooner or later; nevertheless, by means of absolute rest, energetic antiphlogistic treatment, ice, venesection, leeches and low diet, as, also, later, by the frequent use of opium, the inflammation may, in some cases, be prevented. We have had some successful cases. There is no certainty, however, even when the superficial wound is nearly healed, and when no synovia escapes on pressure, so that no passive motion should be employed—for fear of anchylosis,—as chronic inflammation may readily occur, the wound re-open, and the consequent suppuration render amputation necessary.

Bullets coursing under the skin on the inner or outer sides of the joint, causing a so-called "seton-wound," or rebounding or grazing-shots, which destroy the skin, may contuse the capsule, so that the joint is opened when the necrosed cellular tissue is thrown off. Great care must be taken in such cases, to employ local and general abstraction of blood, and ice, so as to try and prevent the progress of the inflammation, as otherwise the sloughing is aided and proceeds deeper. We had such a case in 1850, after the battle of Idstedt, in a Rendsburg hospital. A soldier had been struck by a spent rebounding-shot, on the outside of the knee-joint, a round portion of the skin, of the size of a shilling, was contused and mortified. On this sloughing away the capsule also was found to be affected, as, some days later, a hole, the size of a pin's-head, formed in it,—synovia, and afterwards matter was discharged. We were able, by suitable treatment, and especially by opium, to moderate the severity of the inflammation so much, that the process was very tardy. Although, however, the wound in the capsule had been apparently healed for a long time, yet, little by little, all the symptoms took place of chronic scrofulous inflammation of the knee-joint, deposits of pus occurred between the muscles of the thigh and leg, and the loss of secretions was so extensive, that only removal of the limb seemed likely to save the patient, although anchylosis had already commenced. I, therefore, performed amputation of the thigh on the 10th of November, nearly sixteen weeks after the injury,—the patient died eight days later of pyæmia.

Of course, injury of the extension of the capsule in front of the thigh under the rectus, is to be so considered, although it has frequently been described as a separate bursa. We have had 3 such cases, 2 of them recovered by cautious treatment. In the third case the bullet had entered the middle of the thigh, and was cut out of this extension of the capsule; at a later period, from purulent synovia flowing between the muscles of the thigh, an extensive inflammatory infiltration was set up,—the surgeon in attendance supposed this to be traumatic erysipelas; and, in spite of Dr. Stromeyer's advice, made a long incision through the skin of the whole thigh; by this the soft parts became so degenerated, that, although amputation was later performed the life of the patient could not be saved.

In most injuries of the knee-joint, the bone is, likewise, more or less affected, and the diagnosis is easier, or more difficult, according to the extent and part injured. If the bullet has entered anteriorly or laterally, or, if one of the portions of bone is crushed, the finger easily discovers the whole extent of the injury. If it enters by the popliteal space, or close before the hamstring tendons, the knee being strongly flexed, it is frequently very difficult at once to form a thorough diagnosis. However, when the bone is extensively comminuted, the swelling and tenderness of the limb become, in general, so considerable soon after the injury, that it is difficult to misinterpret the serious nature of the case, even when it is not possible to reach the wounded spot by the finger. If a bullet has entered the joint, and can be felt externally, as occasionally occurs, it must not be supposed the bone has escaped injury, for even a spent-ball, with just sufficient strength left to pierce the integument and capsule, also bruises the bone severely, for it must impinge upon it. The consequence is, that extravasation of blood occurs in its spongy substance; I have already shown of what importance this infiltration becomes in injuries of the epiphyses.

If the shaft of the femur or tibia is comminuted or penetrated by a bullet some inches away from the joint-ends of these bones, it is frequently impossible to determine at once, whether the joint is affected or not. In younger individuals, it may be always hoped that the fissuring has not proceeded farther than the boundary between the epiphysis and shaft; we have frequently observed wounds of this kind, where no inflammation of the joint has occurred and the patient has fully recovered with complete mobility of the knee-joint.

All gun-shot injuries of the knee-joint, in which the epiphysis of the femur or tibia has been affected, demand immediate amputation of the thigh. It is a rule of deplorable necessity, already given by the best authorities, and which our experience fully confirms. In vain have we often made the attempt to leave the case to nature, to save an unhappy man the loss of his limb, on account of a slight injury, but, just so often have we had cause to repent that amputation had not been performed in the first instance. The symptoms of absorption of pus occur occasionally in such cases so rapidly, and with such severity, that all help is in vain. One may, indeed, retard the occurrence of suppuration and moderate its severity so much by the strictest antiphlogistics, that, at first, the progress of such a case, may appear to give the best promise, but, little by little, the process extends to the whole joint; the suppuration breaks through the capsule above and below; the thigh and leg are traversed by purulent sinuses, and even amputation, undertaken at this late period, affords little hope of saving life. Larry, indeed, "Chirurgische Klinik," states that, "he healed numerous cases, even when the bone was injured, by means of quiet position, strict antiphlogistic treatment, pressing out and drawing out by a syringe [?] the fluid from the joint, complete closure of the wound, &c." However, he remarks in the same work, "In wounds of the knee-joint, with comminution of the femur, amputation must be performed." In the same way, Guthrie states—"Gun-shot Wounds of the Extremities:—Every gun-shot wound of the knee-joint, when one or both epiphyses are struck, require immediate amputation. He has not seen a single case recover without removal of the limb." Further on he says—"In slight wounds of the patella, the attempt may be made to save the limb, so, also, if only the capsule is injured, and this to a slight degree. In such cases, the wound of the capsule sometimes heals quickly, and yet amputation is later demanded." He considers the strongest antiphlogistics required in all such cases, and poultices the surest means to obviate the necessity of amputation. We can well subscribe to the opinions of this excellent author.

We have also observed that injuries of the patella may give good hope of preserving the limb. In the year 1848, a Prussian soldier recovered with anchylosis of the knee-joint, in a Flensburg hospital, in whom the bullet had entirely comminuted the patella, and who expressed himself with the strongest determination against amputation of the thigh. In the year 1850, we had a remarkable case of injury of the patella, in which life and limb were similarly preserved. A conical bullet had struck the anterior surface of the patella obliquely, and caused a prolonged lacerated wound, as that of a grazing-shot. The surgeon in attendance, judging from the appearance of the wound, had made no special examination, and the external wound soon healed by simple cold-water dressing. Little by little, however, chronic inflammation came on, with dropsical distension of the joint, and gradual weakening of its ligamentous structures, so that, at last, by contraction of the flexor muscles, incomplete luxation of the tibia flexed on the thigh took place backwards. It was later found that a piece of the bullet was lodging in the patella, it was removed with some difficulty by incision. The bullet had evidently been split, after it had penetrated the patella, the larger portion had then passed on, as often observed in the cranial bones. The flexion, with partial luxation of the knee was gradually removed by Stromeyer's Extension-machine. Hennen, also, relates, in his "Principles of Military Surgery," two cases of recovery after comminution of the patella. In one of them he had bled the patient, in all, to 235 ounces of blood, and had ordered him so little diet, that he thinks, in comparison, Valsalva's dietary was immoderate.

Amputation of the thigh was so fatal in our campaigns, that, latterly we were most unwilling to perform this operation. Of 128 patients, in whom the thigh was amputated, in the three campaigns, 77 died, but 51 recovered. Almost all those operations proved fatal, in which a considerable infiltration of the soft parts was already present at the operation,—the number of these was not small, as the wounded were, partly, brought to the hospital by long and bad roads, partly, such a number of patients were at once brought in after the greater battles, that many of the necessary operations could not be performed within the first twenty-four hours. In such cases we have often said to ourselves, that amputation was, indeed, the only means of saving life, and that even this cruel means could afford but slight hope.

The consequences were especially unfavorable in hospitals, which were necessarily overfilled with severely wounded patients after a great battle. Hence, in cases of extensive comminution of the knee-joint, we could not hope, by means of antiphlogistic treatment, to preserve the life of the patient sufficiently long, until after the occurrence of laudable suppuration, amputation might promise a better result. Stromeyer sought this in some cases, when it was already too late for primary amputation, by opening the capsule freely on each side, so that the pus might have free exit. One of these cases was in a young man of strong constitution, who had the external condyle of the femur comminuted, in the battle of Fridericia, by a bullet, and in whom it was at first doubtful whether the joint was injured. However, about the seventh day, as threatening symptoms of injury of the joint occurred, Dr. Stromeyer laid open the whole wound-canal, extracted all loose splinters, and incised the joint freely on either side; the copious secretions flowed freely from these openings, and the state of the patient was every thing that could be wished. Suppuration soon diminished, the wound had a good appearance, and we hoped soon for recovery with anchylosis of the joint,—when, on the 14th day after the injury, the patient was attacked by a violent rigor, which was soon followed by the other symptoms of pyæmia. The patient died three weeks later, when, according to our opinion, he could not have lived eight days, had his leg been amputated on the first recognition of the injury.

We did not, at first, attempt a proper resection of the knee-joint because, according to the experience of other surgeons, we considered this operation as still more dangerous than the amputation of the thigh. We feared also the extensive surface of spongy tissue which would remain in the wound, after removal by the saw of the epiphyses of the femur and tibia. The first results of opening the capsule in the case just mentioned, encouraged us, at the end of the second campaign to attempt resection in the next suitable case, when such considerable infiltration had been caused by an injury of this sort, that amputation of the thigh could offer no chance of life. For the later usefulness of the limb, it seemed to us decidedly necessary that anchylosis of the limb should result, and that no great portion of the bone should be removed. Such cases, therefore, where either the bullet had entered the joint and slightly injured the bone, or where, besides an injury of the capsule, merely a slight contusion or grazing of the bone was present, seemed to us alone suitable for the operation; for, where the extent of the injury to the bone had rendered it necessary to remove considerable portions of one or of both bones—one could not expect to leave the patient even a partially useful limb.

It was only towards the end of the third campaign that a case occurred, which appeared to us thoroughly suitable for attempting resection of the knee-joint.

The volunteer B——, lance-corporal in the 2nd Schleswig-Holstein rifle-corps, was shot in the left knee while on a recognizance near the Schley, on the morning of the 31st of December, 1850; he was brought in the evening to the garrison-hospital at Rendsburg. The bullet had entered on the outer side close above the commencement of the cartilaginous investment of the external condyle, and had left a round hole both in the trousers and the skin. When the patient came into the hospital, the joint was already distended, and great pain was experienced on the slightest movement of the leg. The surgeons who first examined the case did not agree as to whether the joint was injured or not. As some bloody serum escaped on pressure, it was asserted that the joint was involved, while others, who had been unable to detect any exposure of the bone by the finger, were of opinion that the fluid escaped from the infiltrated soft parts. The patient himself thought that the bullet had been a spent one and had immediately fallen out again. However, some particles of bone felt in the wound were positive evidence of a severe contusion of the condyle. It was agreed to put the limb in an easy position, to employ cold applications, and to administer a dose of morphia.

When I saw the patient, the next morning, the knee-joint was much swollen, there was considerable fever and great pain, so that little sleep had been obtained during the night. Nothing escaped from the wound by pressure on either side of the patella. However, I was satisfied that the capsule and the epiphysis of the femur were injured, the evidences being, the important general symptoms, the tenderness and distension of the joint and its neighbourhood, and, on the other hand, the particles of bone in the wound and the round apertures in the integument and clothes. The inability to discover exposed bone by the finger, was explicable on the supposition that the fascia or biceps tendon had slid over the spot. We did not flex the leg for the purpose of examining the wound, on account of the great pain experienced by the patient. I supposed the external condyle to be comminuted with a fissure extending to the joint, or that the bullet was lodged in the condyle. On account of the bad results of amputation of the thigh, after the first twenty-four hours, and with such an infiltration of the soft parts, we were not encouraged to perform that operation; had it been proper, it should have been performed the evening previously, as every hour's delay rendered the prognosis far worse. It seemed to me, therefore, that this was, on all accounts, a proper case for resection. It was determined to wait for the return of the Surgeon-in-Chief, who was then absent, but expected the next day.

Towards evening the fever became very violent, the pain increased more and more, and delirium commenced. After venesection and ice applications, some hours of sleep were obtained. It was evident, the following morning, that the capsule must be injured; bloody synovia, in some quantity, escaped by slight pressure. During the day the pulse became rapid and small. The swelling of the joint, indeed, became no less, though that of the thigh and leg diminished after the escape of the synovia. Twenty leeches were applied and a couple of ice-bladders, so placed that they covered the joint on either side. Hardly any sleep was obtained at night.

Dr. Stromeyer, who returned on the 3rd of January, at once recommended the wound to be dilated longitudinally, on purpose to examine thoroughly the nature of the injury,—if the capsule alone was injured, that the joint should be opened freely on the outer side,—if the bone was involved, that resection should be at once performed.

Dr. Fahle performed the operation under chloroform. On making an incision, four inches long, longitudinally, on the outer side of the limb, through the skin and fascia, a large quantity of purulent fluid escaped from the joint. On slight flexion of the limb, the finger could be passed through the bone, and feel the bullet lying free in the joint. It was evident, therefore, that the injury had occurred when the knee was bent; and when extended, the tendon of the biceps had covered the opening and prevented a thorough examination. A transverse incision, five inches long, was now carried across the patella from the wound, the patella freed by division of the ligamentum patellæ and the quadriceps tendon, and then the joint opened wide by division of the lateral and crucial ligaments. The bullet had only struck off a flat piece of bone of the size of a shilling from the inferior surface of the external condyle, which, with the bullet, lay loose in the joint. A round piece of the trousers was fixed in the spongy tissue of the bone. A piece of one and a half inches long was sawn off the femur, and the semilunar cartilages removed. As the saw had cut obliquely, so that on extension, the posterior part of the sawn surface pressed against the tibia, a further wedge-shaped portion was removed. The extravasation of blood around the injury was well seen on the sawn surface. The leg was extended and the transverse and two ends of the longitudinal incision united by interrupted sutures.

The loss of blood during the operation was not very much. The limb was at once laid, slightly flexed, on a long, padded, wooden splint, and ice applied. The pain soon decreased, and the patient had a good night. The frequency of the pulse diminished, and the general state improved day by day,—a copious serous discharge, not in the least sanious, escaped from the wound.

On the 8th of January there was no fever, the pulse normal. the tongue clean, the whole transverse incision appeared to be united by the first intention, the sutures were still left, as they appeared to cause no irritation.

On the 9th of January the ice was omitted, at the desire of the patient, and the wound, which now secreted a thick yellow pus, was covered with oiled charpie. On removing the sutures on the 10th of January, the outer end of the transverse incision gaped open and allowed a full sight of the joint. On the sawn surface of the femur, good granulations were already present, the cartilaginous investment of the tibia seemed to be yet unchanged.

On the 16th the cartilage began to discolour and soften, and the following day a piece could be removed by the forceps. The granulations were growing from the sawn surface of the femur, and began already to unite with the granulations arising from the lower border of the transverse incision. At this time frequent spasmodic contractions of the muscles of the calf occurred, which were very painful and led to increased flexion of the leg, so that both the edges of the wound and the surfaces of the bones became wider apart from one another. As the painful cramp still became worse and worse, the limb, on the 21st of January, was pillowed round and thus extended—by this means great relief was afforded to the patient.

However, the following day, fresh pain and cramp occurred, the pulse was feverish, the wound had a dry appearance and yielded but a little serous pus, and became covered on the 23rd, with a greyish, fatty-like layer of exudation. Profuse diarrhœa occurred at the same time, and weakened the patient more and more; a considerable accumulation of pus formed between the muscles of the calf, the skin became jaundiced, and the fever very severe. A troublesome cough set in, and the patient died on the 3rd of February, in a soporose condition.

On examination, the whole of the right lung was in grey hepatisation, and full of larger or smaller deposits of pus. At the apices of the lungs were numerous tubercular masses, and in the spleen and kidneys, there were a number of pyæmic abscesses in different stages and of various sizes. An extensive formation of sanious pus was found in the popliteal space under the muscles of the calf, and the popliteal vein, running through it, was blocked up by a discolored clot. The cartilage of the tibia was only partially thrown off, a portion of the same still hung in shreds from the bone. The luxuriant granulations from the femur were shrunk and discolored.

Although this case was fatal, I am satisfied that its course should lead to further trials of this operation. I am convinced, that the patient would have died within eight days, had the leg been amputated, or the case left to nature, on the 3rd of January. Unfortunately other circumstances were against a favorable result, firstly, the tuberculous constitution of the patient, as evidenced by the state of the lungs: again the bad nature of the air in that hospital, for, at that time in the same place, many other patients, with far slighter wounds, died of pyæmia. We were prevented from repeating the operation by the cessation of hostilities: later observers must determine the indications for resection, or for amputation, after injury of the knee-joint.

*List of Resections of the Elbow performed in the Schleswig-Holstein Campaigns.*

| | Nature of Injury. | Resected Parts. | Place and Day of Injury. | Between Injury & Operation. | Operator. | Method employed | Results. | Remarks. |
|---|---|---|---|---|---|---|---|---|
| 1. v. Schw., Prussian captain. | Comminuted right external condyle, joint opened. | Comminuted condyle and bullet extracted. | Schleswig, Apr. 23, 1848. | 21 days. | Dr. Langenbeck | Chassaignac's. | Cured. | |
| 2. v. W., Danish major. | Comminuted left external condyle. | The three joint ends. | Oeversee 24 April 1848 | 19 do. | Ditto | Langenbeck's. | Ditto | Mobility of elbow. |
| 3. K., private of Hanover. | Comminuted end of right ulna. | The three joint ends. | Ulderup, 6th April, 1849. | 1 do. | Dr. Stromeyer. | Jaeger's. | Ditto | |
| 4. v. M., Saxon Lieutenant. | Comminuted end of left ulna in 18 pieces. | Ends of ulna and radius. | Dueppel, 16th Apr., 1840. | 8 do. | Ditto | Liston's. | Ditto | Serving again. Slight mobility. |
| 5. R,. Saxon rifleman. | Comminuted end of left humerus. | End of humerus and the olecranon. | " | 2 do. | Dr. Langenbeck. | Langenbeck's. | Death May 14, 1849. | Hæmorrhage. Amputation. 1 May. |
| 6. Br. S.-Holstein private. | Comminuted end of right ulna. | End of ulna and radius. | Kolding, April 23, 1849. | 2 do. | Dr. Stromeyer. | Jaeger's. | Cure. | Extensive mobility. |

| | Nature of Injury. | Resected Parts. | Place and Day of Injury. | Between Injury & Operation. | Operator. | Method employed. | Result. | Remarks. |
|---|---|---|---|---|---|---|---|---|
| 7. M. v. F., acting-officer Schles.-Hol. | Splintering of right internal condyle. | Epiphysis of humerus and the olecranon. | Kolding, Apr. 23, 1849. | 3 days. | Esmarch. | Liston's. | Cure. | Gangrene of fore-arm. |
| 8. Br. S., Schles.-Hol. private. | Comminuted end of left humerus. | Two inches of end of humerus and the olecranon. | ,, | 3 do. | H. Schwartz. | Jaeger's. | ,, | Extensive mobility. |
| 9. Chr., Schles.-Hol. private. | Comminuted end of left humerus. | Two and a half inches of end of humerus and the olecranon. | ,, | 3 do. | Dr. Stromeyer. | Liston's. | ,, | Incomplete anchylosis. |
| 10. H., Schles.-Hol. private. | Splintering of right olecranon. | The three joint ends. | ,, | 23 do. | H. Schwartz. | ,, | ,, | Slight mobility. |
| 11. Soe., Danish private. | Comminuted lower third left humerus. Fissures to joint. | Four inches of end of humerus. | ,, | 4 do. | Dr. Niese. | Langenbeck's. | ,, | Incomplete anchylosis. |
| 12. Br., Schles.-Hol. private. | Coronoid process of right ulna struck off. | The three joint ends. | ,, | 37 do. | Hausen. | Jaeger's. | Death, July 11, 1849. | Tubercles, lungs and mesenteric glands. |
| 13. S., S.-Holstein ensign. | Comminuted head of left radius. | Head of radius and one and a half inches of ulna. | Fridericia, July 6, 1849. | 1 do. | Esmarch. | Liston's. | Cure. | Perfect anchylosis. Serving again. |
| 14. O., S.-Holstein private. | Comminuted head of right radius and ulna. | Head of radius and three and a half inches of ulna. | ,, | 1 do. | H. Schwartz. | ,, | Death, July 10, 1849. | Phlebitis. Pyæmia. |

| | Nature of Injury. | Resected Parts. | Place and Day of Injury. | Between Injury & Operation. | Operator. | Method employed | Results. | Remarks. |
|---|---|---|---|---|---|---|---|---|
| 15. R., S.-Holstein musician. | Comminution of end of right humerus. | Two inches of end of humerus. | Fridericia, July 6, 1849. | 1 day. | Dr. Stromeyer. | Guepratte's. | Not healed. | Necrosis of humerus. |
| 16. Kr., S.-Holstein private. | Comminuted end of left humerus. | One inch of end of humerus. | ,, | 2 do. | ,, | Liston's. | Cure. | Perfect anchylosis. |
| 17. W., S.-Holstein private. | Comminuted external condyle and head of radius. | The three joint-ends. | ,, | 2 do. | Esmarch. | ,, | Ditto | Incomplete anchylosis. |
| 18. S., S.-Holstein private. | Right olecranon struck off. | Ends of radius and ulna. | ,, | 2 do. | Marcus. | H-incision. | Ditto | ,, |
| 19. Kr., S., Holstein private. | Splintering of end of humerus and of olecranon. | Ends of humerus and ulna. | ,, | 2 do. | Goetze. | H-incision. | Death 30 July, 1849. | Phlebitis in bone. Pyæmia. |
| 20. H., S.-Holstein private. | Right olecranon fissured. | Olecranon and end of humerus. | ,, | 3 do. | Dr. Stromeyer. | Liston's. | Cure. | Anchylosis. |
| 21. B., S.-Holstein serjeant. | Comminuted end of left humerus. | One and three-quarter inches of humerus and olecranon. | ,, | ,, | Dohrn. | Jaeger's. | Death 12 Aug., 1849. | Drunkard. Crural Phlebitis. Pyæmia. |
| 22. R., S.-Holstein private. | Left coronoid process struck off. Radius grazed. | The three joint ends. | ,, | 16 do. | H. Schwartz. | H-incision. | Cure. | Mobility. |

| | Nature of Injury. | Resected Parts. | Place and Day of Injury. | Between Injury & Operation. | Operator. | Method employed. | Result. | Remarks, |
|---|---|---|---|---|---|---|---|---|
| 23 E. S.-private. | Left olecranon splintered. | The three joint ends. | Fridericia, July 6, 1849. | 24 days. | Goetze. | Liston's. | Cure. | Nearly firm anchylosis. |
| 24. G., S.-Holstein dragoon. | Comminuted end of right humerus and head of radius. | Three inches of humerus and end of radius. | Idstedt, 26 July, 1850. | 1 do. | H. Schwartz. | ,, | ,, | Mobility.. |
| 25. H., S.-Holstein private. | Comminuted left olecranon. | Olecranon. | ,, | 2 do. | Esmarch. | ,, | ,, | Anchylosis. |
| 26. A., S.-Holstein private. | Comminuted end of left humerus. | Two inches of humerus. | ,, | 2 do. | Dr. Stromeyer. | ,, | Death later. | Amputation by Danish surgeons. |
| 27. G., S.-Holstein rifleman. | Comminuted end of right humerus. | One inch of humerus and the olecranon. | ,, | 2 do. | ,, | ,, | Cure. | Anchylosis. |
| 28. R., S.-Holstein rifleman. | Splintering of left olecranon. | The three joint ends. | ,, | 25 do. | H. Schwartz. | ,, | ,, | Nearly firm anchylosis. |
| 29. H. S.-Holstein private. | Comminuted end of right ulna. | Ends of radius and ulna. | ,, | 1 do. | Dohrn. | ,, | ,, | ,, |
| 30. Br., S.-H rifleman. | Right internal cond. struck off. Joint opened. | One and a quarter inch of humerus & half of olecranon. | ,, | 1 do. | Bartles. | ,, | ,, | Anchylosis. |

| | Nature of Injury. | Resected Parts. | Place and Day of Injury. | Between Injury and Operation. | Operator. | Method employed. | Result. | Remarks. |
|---|---|---|---|---|---|---|---|---|
| 31. H., S.-H. rifleman. | Com. end of radius and humerus. Olec. bruised. | The three joint ends. | Idstedt, 25th July, 1850. | 3 Days. | Dohrn. | Liston's. | Cure. | Great mobility. |
| 32. Ka.,S.-H. rifleman. | Comminuted end of left humerus. | Four & a half inches of humerus & half olec. | ,, | 3 do. | Bartels. | ,, | ,, | Great mobility. |
| 33. Koc, S.-Hols. rifleman. | Splintered end of the ulna. | Two inches of ulna and head of radius. | ,, | 3 do. | Dohrn. | ,, | ,, | Slight mobility. |
| 34. B., S. H.-rifleman. | Comminuted end of right humerus. | Two inches of humerus. | ,, | 1 do. | Dr. Niese. | ,, | ,, | Great mobility. |
| 35. S.. working tanner. | Comminuted end of right humerus. | Two and a half inches of humerus. | Rendsburg, August 7. | 10 do. | Kunckel. | ,, | ,, | Slight mobility. |
| 36. H., S.-H. private. | Comminuted ends of left radius and ulna. | $2\frac{1}{2}$ inches of ulna and 3 inches of radius. | Missunde Sept. 12th. | 1 do. | Herrich. | ,, | ,, | Mobility. |
| 37. L., S.-H., private. | Com. on right side, external condyle bruised. | Four inches of radius and ulna. | Friedrichstadt, Oct. 4th. | 2 do. | Dohrn. | ,, | ,, | Great mobility. |
| 38. H., S.-H. rifleman. | Comminuted end of right ulna. | Four inches of ulna and head of radius. | ,, | 1 do. | Dr. Stromeyer. | ,, | ,, | Great mobility. |
| 39. Kl., S.-H. private | Comminuted end of left humerus. | Two inches of humerus and the olecranon. | ,, | 3 do. | Herrich. | ,, | ,, | Mobility. |
| 40. W., S.-H. private. | Comminuted end of left humerus. | One and a half inches of humerus. | ,, | 1 do. | Goetze. | ,, | ,, | Mobility. |

*List of Resections of the Shoulder performed in the Schleswig-Holstein Campaigns.*

| | Nature of Injury. | Length of Resected Parts. | Place and Day of Injury. | Between Injury and Operation. | Operator. | Method employed. | Result. | Remarks. |
|---|---|---|---|---|---|---|---|---|
| 1. O., Prussian grenadier. | Comminuted left os humeri by bullet. | $4\frac{1}{2}$ inches. | Schleswig, 23 April, 1849. | 17 Days. | Dr. Langenbeck. | Langenbeck's. | Cure. | |
| 2. Z., Prussian Lieutenant. | Comminuted by traversing bullet. | $2\frac{1}{2}$ inches. | ,, | 19 do. | ,, | ,, | Ditto | Serving again. |
| 3. Sch., Prussian private. | Com. by shot through the arm. | 5 ditto. | ,, | 25 do. | ,, | ,, | Ditto | |
| 4. G., Hanover private. | Comminuted left os humeri. | $2\frac{3}{4}$ ditto. | Dueppel, 5 June. | 2 do. | ,, | ,, | Ditto | |
| 5. L., Danish dragoon. | Comminuted on right and glenoid cavity. | 3 ditto. | Schleswig, 23 April. | 17 do. | Lauer. | ,, | Ditto | |
| 6. H., Danish rifleman. | Comminuted left os humeri by bullet. | 4 ditto. | Oeversee, 24 April. | 21 do. | Dr. Langenbeck. | ,, | Ditto | |
| 7. K., Danish private. | Ditto. | 3 ditto. | Schleswig, 23 April. | 4 do. | Esmarch sen. | ,, | Death, April 29. | Pyæmia. |
| 8. St., Hanover rifleman. | Comminuted right os humeri by bullet. | 3 ditto. | Ulderup, April 6th 1849. | 1 do. | Callisen. | ,, | Death, M. 30, 49. | Hæmor. ligat. of axil. and subclav. |
| 9. S., S.-Hol. private. | Right os humeri struck off. | $2\frac{1}{2}$ ditto. | Fridericia, 6th July. | 1 do. | Weber. | ,, | Cure. | |

| | Nature of Injury. | Length of Resected Parts. | Place and Day of Injury. | Between Injury and Operation. | Operator. | Method. employed. | Result. | Remarks. |
|---|---|---|---|---|---|---|---|---|
| 10. B., S.-Hol. private. | Left os humeri grazed. Spina scapulæ commd. | $1\frac{1}{2}$ inches. | Fridericia, July 6th. | 4 Days. | Goetze. | Langen-beck's. | Death, 22 July 1849. | Hæmorrhage. Pyæmia. |
| 11. L., S.-Hol. rifleman. | Right os humeri grazed. Later partial necrosis. | 3 ditto. | ,, | 35 do | Francke. | ,, | Cure. | |
| 12. E., S.-Hol. private. | Ditto on left. Later partial necrosis. | 2 ditto. | ,, | 35 do. | Hermann Schwartz. | Dr. Stro-meyer's. | ,, | |
| 13. K., S.-H. private. | Comminuted left os humeri. | 3 ditto. | Idstedt, July 25, 1850. | 24 do. | H. Schwartz. | ,, | Death Mid. Sept. | Hæmorrhage. |
| 14. H., Danish private. | Comminution with perforating wound of chest. | 3 ditto. | ,, | 18 do. | Francke. | Langen-beck's. | Death, 25 August. | Hæmorrhage. Bullet in thorax. |
| 15. L., S.-Hol. rifleman. | Os humeri struck off. | 5 ditto. | Missunde, 12th Sept. | 1 do. | Esmarch. | Francke's. | Cure. | |
| 16. Schm., S.-H. rifleman. | Comminuted left os humeri. | $2\frac{1}{2}$ ditto. | Friedrichstadt, 4th Oct. | 1 do. | Herrich. | Langen-beck's. | Death, 1 Nov. | Pyæmia. |
| 17. K., S.-Hol. rifleman. | Ditto right. | 4 ditto. | ,, | 1 do. | Dohrn. | Francke's. | Cure. | |
| 18. H., S.-Hol. rifleman. | Ditto ditto. | 2 ditto. | ,, | 1 do. | Francke. | ,, | ,, | |
| 19. L., S.-Hol. rifleman. | Ditto left. | 5 ditto. | ,, | 14 do. | Thiersch. | Dr. Stro-meyer's. | Death, 27 Oct. | Hæmorrhage. Pyæmia. |

CASES OF

# RESECTION

IN

CIVIL PRACTICE

ON

TONIC TREATMENT THROUGHOUT.

BY

S. F. STATHAM.

ASSISTANT SURGEON TO THE LONDON UNIVERSITY COLLEGE HOSPITAL.

---

LONDON:

J. W. TATTON, 29, QUEEN ANNE STREET.

1856.

# CASES OF RESECTION

ON

# TONIC TREATMENT THROUGHOUT.

The Translator, who was in service at the termination of the last Schleswig-Holstein campaign, here ventures to give the Cases of Resection which he has performed since that time. Before entering upon these few cases, some remarks may be made on the preceding pages.

It is impossible for me to agree with the authors on the antiphlogistic treatment, so strongly recommended, on the occurrence of diffuse inflammation. It is easy to be proved, experimentally, that no diffuse inflammation will spread towards the heart, if the patient is put on the best tonic treatment, at the same time that the inflammation is treated antiphlogistically. It may, at the same time, spread somewhat distally.

The knowledge of this fact will, I am satisfied, answer the end which Dr Stromeyer seeks, for closing the mouths of veins in a sawn bone. (p. 6.)

Several resections were not tried, to my knowledge,—such as resection of the wrist or ancle,—in one case this operation was talked of for a joint of the lower jaw on one side, and was certainly indicated, by the fear of approaching anchylosis. Dr. Stromeyer's operation for resection of the shoulder-joint, cannot be recommended, as the transverse incision is very apt to gape, and that, even to the extent of three inches, so that, of course in such cases the sutures are of no use, or serrate the skin.

The more frequent fatal termination of wounds of the left shoulder (page 68), seems to me easily explicable, from that joint being before the body, both in the act of firing and when using the bayonet. The impinging of the bullet so frequently "on the right side, on the tendon of the biceps," seems to prove the same point.

The excellent pathological observations made in this translation are well worthy of remark—such as, on the nature of sequestra. Again, such facts, as that the truncated end of the humerus unites with any portion of the anterior costa of the scapula.

---

Three of the resections are of the ancle-joint, two of the knee, and one of the elbow. It may be said, and justly so, that all were considered to be cases requiring amputation.

Case I.—Resection of Scrofulous Astragalus (from Vol. XXXVII. of the "Medico-Chirurgical Transactions").—Henry Cudden, æt. 5, of strumous tendency, was said to have had weakness of the left ancle since birth. At Christmas, 1851, a swelling appeared below the outer side of the left ancle, which was blistered; since May he has been under hospital treatment; painting with iodine was frequently employed; latterly the formation of matter pointed naturally on the inner side, and required opening outside the joint.

August 25, 1852.—The integument was much diseased about the ancle; but on closer examination, and after a week's rest in the hospital, it was found to be actually implicated only where corresponding to the situation of the astragalus. Chloroform being administered, the fistulæ were thoroughly examined. On the outer side the probe reached the surface of the astragalus, which was exposed and softened, and there was a fistula leading backwards by the side of the calcaneum; on the inner side a probe passed easily along the posterior face of the astragalus. The ancle-joint was healthy.

Medicines and local applications having been fairly tried without avail, amputation below the knee would probably have become inevitable; and as his health was already materially suffering, resection of the astragalus, and of any portion of the calcaneum that might be necessary was considered justifiable, if only to be followed subsequently by removal of the limb.

August 27.—My friends Messrs. Marshall and Clover assisting me, an incision, three inches long, was carried along the outer side of the extensor tendons of the toes, and another to fall into the middle of this one from the outer side of the foot. The finger found carious disease of the neighbouring surfaces of the astralagus and calcaneum. Having lifted up the flaps of the soft parts, and separated the tendons and vessels in front of the joint in one mass from the bone, it was sought to release its head, which proving troublesome, all difficulty was at once removed by cutting through the neck of the astragalus with the scalpel, and then by means of the fingers and sequestrum-forceps, the pieces were dragged out, while the knife freed them from the surrounding parts. During extraction, the posterior portion of the upper cartilaginous surface became separated from the body of the bone, and was removed later; this circumstance much facilitated the operation. The upper surface of the calcaneum, for its posterior two thirds, was found to be carious, and was therefore gouged off to a depth of about one-eighth of an inch. The foot hung perfectly loose; three fingers could be easily introduced to the bottom of the wound, the surfaces of the tibia and fibula were sound the remaining portions of the tarsus offered no reasons for interfering with them. It was found that the tendons of the peroneus brevis and external tendon of the extensor of the toes had been divided, the lateral ligaments to the calcaneum had escaped, and no vessels required ligature, the profuse hæmorrhage being readily checked by cold water. Lint was introduced into the cavity of the wound. The same evening a splint was applied on the inner side of the leg and foot, and a piece of wet lint laid over the wound; this was still large and gaping, as the calcaneum would not enter between the malleoli. The splint and pad were perforated to allow the escape of any wound-secretions.

On examination of the bone, the upper articular cartilage appeared to be unaffected, but easily separated from the carious body of the bone beneath. The posterior articulation of the astragalus with the calcaneum had disappeared. The head and neck of the bone appeared to be sound.

September 1.—Wound suppurating, health fair. Was put on iron and nitric acid, later porter: fish, &c. The foot was never removed from the splint, nor the watery pus from the cavity of the wound, otherwise than by trickling water over it, for a whole fortnight.

At the expiration of this time, September 11th, chloroform being given a third time, the side-splint was changed for one of tin fitted to the back of the leg and foot. The foot was found slightly raised on the inner side, otherwise in good position; the wound was filled from the bottom and sides by coarse vascular granulations, not thoroughly united, so that three passages, admitting a probe loosely, ran to the posterior inner corner of the wound, where a small portion of the calcaneum was exposed (having apparently escaped the gouge); all other parts of the wound presented to the probe a softish mass, which it was not attempted to penetrate. The edges began to draw in and cicatrize, and their neighbourhood became much improved on the state prior to the operation.

About October 10th, Mr. Erichsen examined the wound, and found no bone exposed.

October 15.—By Mr. Erichsen's advice, ointment of the nitric-oxide of mercury was used to the edges of the flabby wound with advantage. The anterior fistula healed, the posterior one became quiescent. A dextrine bandage was applied, and the patient discharged.

November 15.—The wounds were fairly healed.

Christmas and Lady-day.—He is going on thoroughly well, can walk without pain; there is free mobility of the new joint, the cicatrix is becoming much firmer and smaller. Till now a splint has been constantly employed; he may have a boot fitted. Slight inversion of the foot continues, and the leg is about one inch shorter than the other.

June 14.—Mr. Gray, of Cork Street, made him a well fitted boot, the heel raised, an iron support up to the knee (jointed opposite the ancle to allow limited motion), with a broad strap around the ancle, and another band below the knee. The lad is able to walk and run without any pain, and with merely a halt, partly due to the incumbrance of the instrument. The foot is perfectly sound; he can extend it well, flexion of it on the leg is not so easy, the present relative position of the parts being more disadvantageous for this action than before; its mobility is complete. Inversion of the foot has disappeared.

I must acknowledge my thanks to Mr. James Turle, the house-surgeon, for the great care and ability with which he treated the patient.

March 1856. The arch raised since the operation is now normal, and the cure perfect. A model of these feet taken some twelve months after the operation is in the College of Surgeons, London.

CASE II.—Resection of the right Knee-joint.—On my taking charge of patients at the North London Hospital in the summer of 1854, at the beginning of August, one of them a female, H. C., æt 20, was found to have the right knee much distended with matter. Two lateral incisions were at once made after the employment of an exploratory puncture. A large quantity of matter escaped from the cavity of the joint; this did not appear as yet to be more than inflamed. Great relief was obtained and everything went on well for about a week or ten days, when it was found that, after the occurrence of some diffuse inflammation, very extensive sinuses had formed close to the bone in all directions around the joint. It was impossible to ascertain their extent without an operation. Slight but perfectly distinct grating was felt between the patella and femur.

The patient becoming hectic and the soft parts around the joint being in a very bad condition, I determined at once to perform resection, as otherwise the sinuses could not be reached, and as their close proximity to the joint, and as the inflammation which caused them being of low character, would have rendered it very doubtful if they could have occurred—or would be recovered from—without inducing caries of the bones of the joint; again, as such an operation would prevent the necessity of amputation and no other means was sufficient for that purpose.

Amputation was recommended to me at the time of operation.

August 26th.—Four weeks after the joint had been first opened, and about two months after the first occurrence of inflammation in the joint, the operation was performed by means of lateral incisions; this is, I believe, the best method and to have been first employed in this case.

Perpendicular incisions are first made on one side of the joint, of about sufficient extent to thrust out the joint-end, and somewhat beyond the spot where it is supposed the bone or bones will be truncated,—below a horizontal incision connects them. By running the back of the knife from the first perpendicular incisions over the joint to its other side, it is easy to make a flep there corresponding to the first.

This flap is only through the skin, and the flaps are now reflected on either side.

The joint is now opened by dividing the lateral ligaments, and then everything separated from any bone that requires resection of its extremity. By this means no vessel of consequence is divided, the flaps fall naturally into place afterwards—require and have no sutures—allow free discharge of matter, and never let any bone be exposed to the action of external agents.

I thought of this method of operating by having seen the anterior flap retract, and leave the bones exposed and become necrosed, in the case of resection of the knee, mentioned in Dr. Esmarch's work.

After completing the operation by removal of about an inch of the femur and of the tibia, several cavities were found about the joint—perhaps four, two of which would admit a hand. Equal parts nitric acid and water were well applied to the lining membrane by means of tow on a glass rod.

No hæmorrhage or complication occurred, except slight gangrene of one corner of a flap.

Three or four months later a sinus down the calf of the leg, with sloughing from decubitus, on the sacrum, ancles, hips, &c.,—although the patient lay on a water-bed—led to amputation being talked of, however, 4 grains to the ounce of nitrate of silver was injected, and it healed up the sinus.

The patient went out against my wish, April, 1855; "is very thin, hips and leg healed, ancle and back fast healing, still obliged to lie."

March 14th, 1856.—Is found lying in bed, but gets up six hours a day and gets about with the aid of a chair. Joint is quite healed, not tender, but stiff. No matter escapes, and there is no pain.

Case III.—Resection of the Right Ancle-joint.—A strumous female child, aged 5 years, was under my care, at the North London Hospital, for about a year and a-half, with tenderness and stiffness of the joint and some enlargement of the bones at that spot. As other means produced no good effect, blisters and, I believe, a seton were used. After a short time diffuse inflammation attacked the soft parts. It was evident that the bones were likewise affected, and, therefore, after due examination, excision of the parts was proceeded with.

Aug. 4, 1854.—Incisions, as described in the preceding case, were employed, and the astragalus and then one inch of the tibia and fibula were removed.

All these portions were in far advanced caries and bathed in suppuration.

Nothing particular occurred in the case until about eight months later, when, after probing a single fistula which then remained, a sudden attack of phlebitis occurred on the following day.

On the inner side of the knee, above the joint, there was deep tenderness, and a firm prolonged swelling in the position of the femoral vein. The glands were apparently not affected. The whole leg was extremely swollen with a kind of vesication from the distended vessels.

A purgative or so, porter, and four or five meals of meat daily, with warm lead lotions, relieved the dangerous condition at once, and the swelling in some days. Complete cicatrization took place in about a year. It was longer than usual, as the mother had dressed the wound almost the whole time, and the child could not be kept from walking about.

At present, March 1856, it is a perfect cure; the child requires irons, however, up to the knee, as, from the weight of the body and small surfaces of the truncated bones of the leg, there is a strong tendency of the limb to bend outwards at the new ancle.

This must be considered a good case as the patient is decidedly strumous, especially evidenced by the hydrocephalic condition of the head.

Case IV.—Excision of the Elbow.—H. G., female, æt. 74, came to the North London Hospital, August 1854, with neglected inflammation of the left elbow-joint of three months standing. The patient is healthy, but has a loud mitral murmur.

August 16th.—The operation was performed in the usual way, except that, to make use of a previous incision, the longitudinal one was carried down on the outer side. The olecranon and condyles being carious and bathed in suppuration, were removed with the head of the radius. A little chloroform was given to dull the pain.

Everything proceeded well, except a troublesome hiccup, apparently due to her being confined to bed. After a few months, as the wound did not entirely heal, amputation was talked of, however, by strong acid injections, steel-wine, purgatives and gin, the whole was consolidated in ten weeks—of the last treatment.

April, 1855.—The patient has no pain, can raise her hand to her head without any assistance. The fingers are stiff and probably will remain so at her age.

March 1856.—Every thing is going on well, the cure is entire, the fingers still stiff. Nothing extra is worn on this arm, but a piece of flannel.

Case V.—Resection of Strumous Condyles of Femur.—A female child of about 5 years of age, was brought to the same hospital September 1855, with the ends of tibia and femur enlarged, tender and probable suppuration up the thigh, and in the joint.

After an exploratory puncture, in the theatre and under chloroform, giving evidence of the presence of matter close to the condyle of the femur, resection of the knee was proceeded with by means of lateral flaps.

About two inches of the end of the femur was removed; it was enlarged, softened, and the cartilage absorbed in various distinct spots through its entire thickness between the two condyles,—the deeper soft parts were in gelatinous, grey degeneration. Although the head of the tibia was in primary strumous condition, it was merely gouged to a slight extent, on purpose to see if it could be possible to save it, in our present state of knowledge, without removing it from the body.

Everything went on well, as regards the rapid healing of the wound, but, I believe that the head of the tibia ought still to be excised, and, although amputation might be thought of by some surgeons, I am satisfied that a firm joint might be easily obtained by removal of an inch and a-half of the tibia.

The patient went out some months ago. Her health good, a fistula leading to the joint, the head of the tibia and soft parts over it in strumous condition.

Case VI.—Resection of diseased Astragalus.—Later removal of the Malleoli and Gouging of the Surface of the Tibia.—J. H., æt. about 49, painter by trade, came to me as out-patient at the hospital, in June, 1855. Having seen him several times as an out-patient, he was recommended to come into the hospital, as the swelling and ulcer at its apex, on the inner side of the right ancle gave positive proof of disease of bone. He entered July 21, 1855.

He had sprained his ancle some six years before, having previously had rheumatism in the joint. The ancle was sprained at the sea-side, where he had gone for the benefit of his health.

Under chloroform, diseased bone of the surface of the astragalus was found and resection was determined on and performed on July the 25th.

It was performed in the same way as in the first case mentioned, excepting that knowing the difficulty of effecting it properly, I looked about and found a pair of gas-fitters' pincers, which, by slight alteration, became thoroughly suitable for the purpose.

The head of the bone was clipped off by Liston's forceps, and then extracted by those above-mentioned. Liston's forceps were again used, to split the body of the bone in two, and the pieces then extracted by the forceps described.—a pair of a similar kind has been made for me by Mr. Weedon, of Hart Street. These forceps cannot be too much recommended for operations on the larger bones.

On examination there was a white deposit transversely beneath the joint-surface of the bone of about a quarter of an inch in diameter, with some adhesion from about this point to the tibia. The inner side of the astralagus was diseased.

Slight bilious fever occurred, but was quickly removed; the wound granulated healthily, and was nearly filled in a fortnight. It is worthy of notice that the discharge for some days was watery with flakes, this, no doubt, was the cartilaginous substance.

However, a fistula remained, and, on November 21st, it was determined to remove the malleoli. This was done from the inner side alone, hence up till this time there have been fistulæ on both sides, with occasional discharge of sequestra.

I had wished to remove about an inch from the bones of the leg at the the time of the last operation, but was otherwise advised. It had been recommended to me in the first instance. Of the ultimate recovery I do not doubt, and the present (March 21st), state of his joint appears as if no further operation would be required.

These good results from the operation of resection, viz.: in a patient, with organic disease, of 74 years of age, after the elbow had been excised; in another, of 20 years of age, after resection of the knee, when the joint was in very bad condition, and two cases of diseased ancle—being soundly healed, while the two last cases remain in a doubtful condition, merely because an insufficient quantity of bone was removed, justify me in preferring resection of a whole bone to a partial gouging of it. I have had no other operations of resection of joints.

Mr. Solly has spoken of Resection of the Astralagus, that it "sounds well and reads well," and, apparently, would infer that such might be the object of the operation. However, I must add, that they, also, "get well," which is more than can be said of very numerous cases of gouging of bone. Of course, there is no very extensive disease if, "only a slight fistula remains and the patient goes into the country." For my part, I prefer keeping the patient under observation until deep, firm scars, no pain, and no discharge, give positive testimony that the patient is cured.

In performing the operation, the main points are—to ensure there being no hæmorrhage, that would require ligatures being applied,—to divide no tendons or nerves passing over the joint, to make the flaps in such a way that they will lie naturally in place after the operation without any sutures, and never (Stromeyer) to remove the limb from the splint during the progress of cure.

No matter how large the wound is, it appears, under this method of operating, to be nearly filled up with granulations in 10 or 14 days; but one most important adjuvant remains to be considered:—that, as antiphlogistic treatment is the best possible preparative for the winding-shroud in all extensive operations,—the patients should be fed just as well after the operation as before, and that frequently, they require stimulants to prepare them for the operation, if so, these should be continued after the same.

The only cases, in London Hospital practice, where I find antiphlogistic treatment useful is, in operations on the eye,—such as cataract, or the operation for artificial pupil;—probably it is proper where there is no loss of blood.

If diffuse inflammation comes on, the patient should be at once put on the best stimulant and nourishing treatment, though the inflammation may be at its height, at the same time that abstraction of blood, poultices, &c., are employed locally. Iron, acids, quinine, should be given, to enable the patient to assimilate the food he swallows. Alkalies, I have always found positively injurious, unless when demanded by contemporary Bright's Disease. For this, my friend, Mr. Palmer, of Charlotte Street, first recommended to me tonics for the disease itself, and I have found the alkaline double-preparations of iron and quinine to afford eminent service.

Bright's Disease being very common, this slight notice of a point in its treatment, is here given for the sake of success in operations.

Nothing further will now be said of 'low' or diffuse inflammations, the subject requires a distinct consideration.

# INSTRUMENTS

USED BY

## THE TRANSLATOR.

Steel Probe (not too much hardened) for examining if the bone is softened.

Blunt-pointed Knife (Langenbeck's) for deep operations on bones. The blade is continued down the handle. The edge should be soft rather than hard.

Saw for division of any long bone, without thoroughly exposing it, (Langenbeck's.)

Resection Forceps as already described, p. 118, are of great service during the use of the knife above-mentioned.

*If the above are not kept in Instrument-maker's shops, Mr. Statham would be happy to show them.*

*** All Instruments are of full size, but about 2 inches of the handles of the Forceps and Saw are not shown.

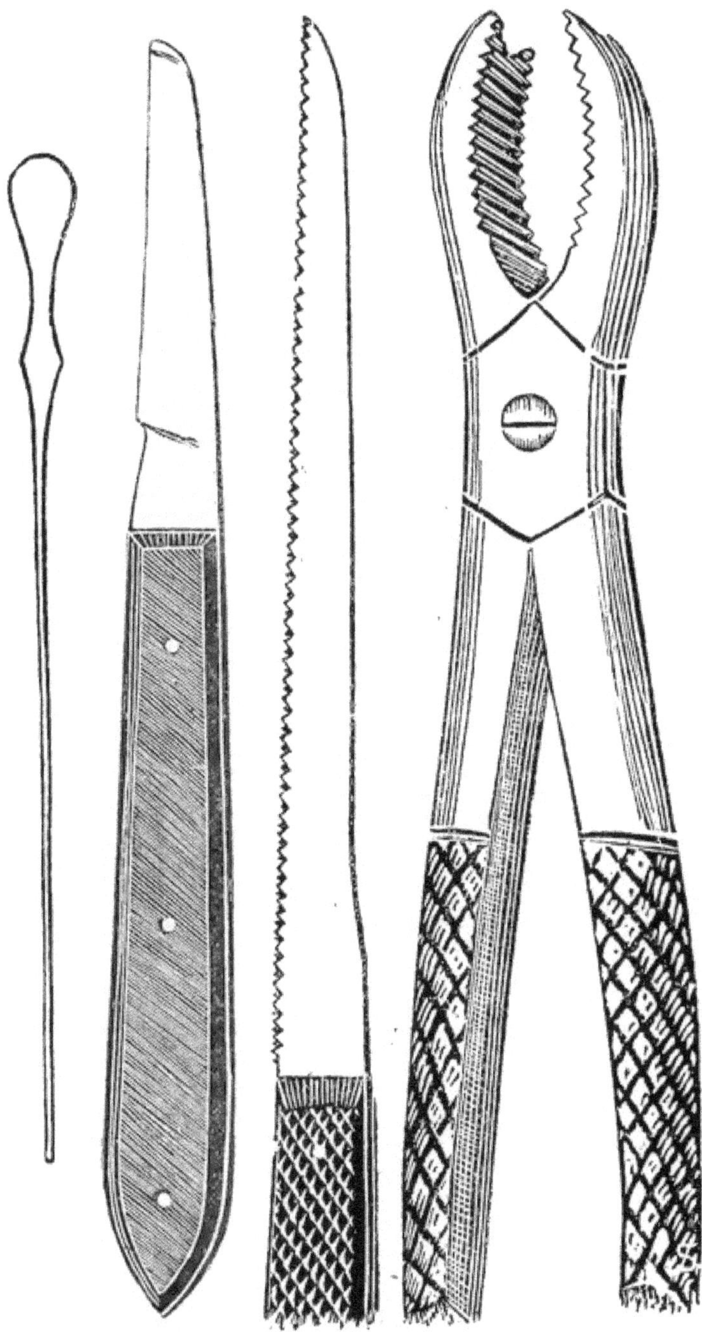

www.ingramcontent.com/pod-product-compliance
Lightning Source LLC
Chambersburg PA
CBHW021818190326
41518CB00007B/651

* 9 7 8 3 3 3 7 1 5 9 7 9 5 *